Complex Systems Studies

WIT_PRESS_

WIT Press publishes leading books in Science and Technology.
Visit our website for the current list of titles.
www.witpress.com

WIT_eLibrary_

Home of the Transactions of the Wessex Institute.
The WIT electronic-library provides the international scientific community with immediate
and permanent access to individual papers presented at WIT conferences.
http://library.witpress.com

Complex Systems Studies

Editors

G. Rzevski
The Open University, UK

C.A. Brebbia
Wessex Institute, UK

WITPRESS Southampton, Boston

Editors:

G. Rzevski
The Open University, UK

C.A. Brebbia
Wessex Institute, UK

Published by

WIT Press
Ashurst Lodge, Ashurst, Southampton, SO40 7AA, UK
Tel: 44 (0) 238 029 3223; Fax: 44 (0) 238 029 2853
E-Mail: witpress@witpress.com
http://www.witpress.com

For USA, Canada and Mexico

Computational Mechanics International Inc
25 Bridge Street, Billerica, MA 01821, USA
Tel: 978 667 5841; Fax: 978 667 7582
E-Mail: infousa@witpress.com
http://www.witpress.com

British Library Cataloguing-in-Publication Data

A Catalogue record for this book is available
from the British Library

Library of Congress Catalog Card Number: 2017958178

ISBN: 978-1-78466-277-6
eISBN: 978-1-78466-278-3

The texts of the papers in this volume were set individually by the authors or under their supervision. Only minor corrections to the text may have been carried out by the publisher.

No responsibility is assumed by the Publisher, the Editors and Authors for any injury and/or damage to persons or property as a matter of products liability, negligence or otherwise, or from any use or operation of any methods, products, instructions or ideas contained in the material herein. The Publisher does not necessarily endorse the ideas held, or views expressed by the Editors or Authors of the material contained in its publications.

Preface

The papers included in this volume describe a range of applications of Complex Systems across different disciplines.

These systems, because of their complexity, require the application of new approaches different from the conventional mathematical models.

Complex issues exhibit some of the following attributes:

- CONNECTIVITY – A system consists of a large number of diverse components, referred to as Agents, which are richly interconnected.

- AUTONOMY – Agents are not centrally controlled; they have a degree of autonomy but their behaviour is always subject to certain laws, rules or norms.

- EMERGENCE – Global behaviour of a complex system emerges from the interaction of agents and is therefore unpredictable but not random; it generally follows discernible patterns.

- NONEQUILIBRIUM – Global behaviour of a complex system is far from equilibrium because frequent occurrences of disruptive events do not allow the system to return to equilibrium between events.

- NONLINEARITY – Relations between agents are nonlinear, which occasionally causes an insignificant input to be amplified into an extreme event.

- SELF-ORGANISATION – A system is capable of self-organizing in response to disruptive events, a feature termed Adaptability. Self-organisation may also be initiated autonomously by the system in response to a perceived need, a feature termed Creativity.

- CO-EVOLUTION – A system irreversibly co-evolves with its environment.

Complex Systems occur in an infinite variety of problems, not only in the realm of physical sciences and engineering, but encompassing fields as diverse as economy, the environment, humanities, social and political sciences.

Some of these applications are described in a number of the papers included in this volume.

The Editors hope that the material printed here will contribute to a better understanding of the way in which complex systems can be solved. They are also grateful to all authors for their excellent contributions and to the reviewers for helping to ensure the quality of this volume.

The Editors

Contents

FATHOMING THE FUTURE OF ARTIFICIALLY INTELLIGENT ROBOTS

BRIAN E. WHITE

Complexity Are Us ← Systems Engineering Strategies, USA.

ABSTRACT

The world abounds with massive efforts to further develop artificial intelligence, mostly with hopes of achieving greater benefits to humankind. Not surprisingly, there is relatively little concern about the dangers associated with the, as yet hypothetical, eventual situation where robots might possess human-like capabilities of cognition, emotional experience, learning, etc. The following five *propositions* will be examined:

1. Could the most advanced robots 'evolve' to truly human-being levels of achievement within a foreseeable time frame?
2. Will robots ultimately take over *all* the jobs, including the making of robots, with relatively few human owners of robots (and most everything else) in charge?
3. Will humans live much longer and essentially turn into pseudo-robots through receiving more replacement body parts, even involving portions of the brain?
4. What are the possibilities of being psychologically manipulated by authoritarians using Big Data in knowing what citizens care about and how people think?
5. Will humans keep losing not only manual jobs but also knowledge positions with increasing robotic capabilities and the attractions of robotic replacements?

The conclusive answers: a 'no' to Proposition 1; a 'maybe' to 2; possible 'yeses' to 3 and 4; and a definite 'yes' to 5; will be explained.

This subject quite obviously overlaps some combination of at least three of our conference themes, viz., Complex Systems Engineering, Global Issues and Social Systems. The paper's focus will be on framing the above-described topics in a matrix with two dimensions: *holistic thinking perspectives* (big picture, operational, functional, structural, generic, continuum, temporal, quantitative and scientific) and *journalist questions* (who, what, where, when, why and how). Several of the more interesting topics from this milieu will be elaborated upon to stimulate further thought, discussion, and research efforts.

Keywords: artificial intelligence, behaviors, complex systems, complex systems engineering, families of robots, global issues, humans, jobs, robots, social systems.

1 INTRODUCTION

The topics of this paper's title and abstract are emergent, rather popular, quite controversial, far-ranging, and very challenging. The author here attempts to treat some of this and the supporting literature in a meaningful fashion to not only express his point of view but also encourage others to weigh in with their perspectives and argue their similar or contrary viewpoints, hopefully with additional research.

The remainder of this introduction includes some relevant acronyms/definitions, the discussion framing, and a natural evolution of robotic phases. Section 2 sets up considerations of the propositions (treated in Section 3), and this author's conclusions are summarized in Section 4. In addition to the referenced citations, this author's comments inserted within quotations are demarcated by brackets, i.e., [...].

© 2018 WIT Press, www.witpress.com
DOI: 10.2495/DNE-V13-N1-1-15

1.1 Acronyms and definitions

The following terms, listed in alphabetical order, will be utilized in the paper.

- Agency: set of ('outside' oriented) mental abilities of thinking and doing including communication, logic, memory, morality, planning, recognition, and self-control [1, p. 11]
- AGI: artificial general intelligence; human-level AI [2, pp. 8, 17]
- AI: artificial intelligence
- ASI: artificial super intelligence; intelligence greater than humans [2, p. 8]
- Automatic: working by itself with little or no direct human control
- Automation: the use of largely automatic equipment in a system of manufacturing or other production process [3]
- Experience: set of ('inside' oriented) mental capacities of sensing and feeling including consciousness, desire, embarrassment, fear, hunger, joy, pain, personality, pleasure, pride, and rage [1, pp. 10–11]
- Robot: a machine capable of carrying out a complex series of actions automatically; a machine resembling a human being (a humanoid) and able to replicate certain human movements and functions [3]

1.2 Framing the discussion

An overview of the general topic of robots and humans is summarized in Table 1. To better inform others a good journalist [4] and Zachman [5] ask basic questions namely, Who?, What?, Where?, When?, Why? and How? Holistic systems thinking descriptors, Big

Table 1: Contexts and questions regarding the general subject of robots and humans–potential implications.

Question Thinking Perspective	Who?	What?	Where?	When?	Why?	How?
Big Picture	robots and humans	robots replacing humans to various degrees	on Earth primarily	near to far future	because robots can become much more capable	by robots becoming more 'intelligent'
Operational	several robotic and human types	robots mimicking, challeng-ing, or surpassing human behavior	any place not necessar-ily popu-lated by humans	present to far future	to benefit, replace, or control humans	by robots learning skills, getting knowledge, and 'boot-strapping' themselves

(Continued)

Table 1: (*Continued*)

Thinking Perspective / Question	Who?	What?	Where?	When?	Why?	How?
Functional	special-ized robots and cyborgs	specific capabili-ties/ char-acteristics	physical areas of applica-tion	present to near future	to empha-size de-velopment aspects	by building on current research efforts and capabilities
Structural	robotic/ human body parts and intercon-nections	how parts work and are inte-grated	part locations in robots and humans	whenever parts are active	to learn how parts contribute to func-tionality and opera-tions	by experi-menting with parts types and concepts for their intercon-nections
Generic	machines and people	artificial vs. human intelli-gence	wherever intelli-gence is found	whenever intelli-gence is found	to un-derstand promise/ danger of machine intelligence	continued re-search, study, analysis, synthesis, and publication
Continuum	continued evolution of ma-chines and humans	cumulative collection of robotic/ human properties	any-where	anytime	to discover and charac-terize new/ important trends	by maintain-ing general and objective viewpoints
Temporal	ages of robots and humans	rates of change in capabili-ties	wherever there is sig-nificant current focus	whenever there is significant current focus	to measure robotic and human achieve-ment trends	by 'psycho-logically' and physically comparing robots and humans
Quantitative	relative popula-tions of robots and humans	distribu-tion of robotic functions	distribu-tion of robotic locations	distribu-tion of robotic timelines	to measure relative seriousness of robotic activities	by estimat-ing numbers, percentages, and densities
Scientific	robotic research-ers and authors	levels of under-standing of robotic topics	hotbeds of robotic study and discourse	years of robotic technical break-throughs	to get underpin-ning of robotic engineer-ing	review of literature and potential personal interviews

Picture, Operational, Functional, Structural, Generic, Continuum, Temporal, Quantitative, and Scientific categorize the perspectives taken in attempting to answer any of the questions. [6]

The reader should skim through Table 1 to get a sense of where we are going. In exploring any question/descriptor pair, other questions/descriptors may be invoked.

1.3 Natural progression of robotic phases

Suppose there are eight phases in a natural evolutionary progression toward the ultimate sophistication of robots, as depicted in Table 2. Examination and interpretation of this table resonates with many aspects of Table 1. These phases can be viewed in essentially reverse order timewise of the five propositions posed in the abstract. Specifically, Proposition 1 roughly corresponds to phases VII and VIII, 2 to VI, 3 to V, 4 to IV and 5 to I–III. The five propositions are considered in Section 3, drawing upon research and pronouncements of experts in the field.

The next section expresses thoughts that come to mind to help resolve the propositions. Assume several desired (and perhaps undesired) *premises* about what robots might eventually accomplish, i.e. the **What**. Then consider their ramifications by addressing **Who**, **When**, **Why**, **Where**, and **How**.

Table 2: Hypothetical evolution of robotic capabilities.

Phase-Era	Label	Brief Description
I-Past and Present	Automation	What's been going on since at least the times of Henry Ford and the industrial revolution
II-Present	Artificial Intelligence	Research and experimentation since the mid-20th century, especially in the past couple of decades
III-Present and Future	Robots Doing Many Jobs	Robots replacing manual and even white-collar and/or knowledge-based workers, particularly in the past decade; many more jobs are threatened
IV-Very Near Future	Big Data Mind Control	Autocratic psychological influence of many through leveraging huge amounts of automatically collected data plus some misinformation; is already happening!
V-Near to Far Future	Humans Turning into Robots	Seems inevitable as more of us receive artificial body parts/transplants, including portions of the brain; this is bound to greatly increase longevity!
VI-Far to Very Far Future	Robots Doing All Jobs	Robots doing *all* the jobs leaving most humans with no occupational opportunities; the very few elite in control will own everything, including the robots
VII-Very Far to Distant Future	Robots Becoming Human-Like	AI succeeds in creating robots capable of human aspirations and qualities like ambition, cognition, conscience, emotion, faith, learning, memory, morality, reason, religion, and self-awareness
VIII-Very Distant Future	Robots Surpassing Humans	If/when robots achieve Phase VII, they will quickly and irreversibly far surpass humanity and potentially marginalize, minimize, or eliminate us!

2 SOME THOUGHTS REGARDING THE FRAMING QUESTIONS
The basic questions are next addressed in the aforementioned order.

2.1 What?

Here posed are a few *very long term* (see **When**) outcomes in terms of robotic states; presumably, Premises 1 and 2 below might be desired, and Premise 3 is undesired.

Premise 1. All jobs are done by robots including making and repairing robots.

If robots do all the jobs, one wonders how anyone would satisfy their innate human aspirations of doing something useful or pleasurable in life. On the other hand, many people would clearly prefer not to be forced to work in undesirable occupations just to make a living, so that can be an upside. Ever since the US was an agrarian society, and transitioned through the industrial and information ages, we've been worried about automation taking over our jobs. The remedies, of course, as has been espoused many times, are proper education and effective retraining so the affected people can aspire to new job opportunities and actually fulfill them successfully with at least adequate expertise. How strongly do these measures still hold?!

Premise 2. Robots miraculously solve many heretofore intractable mysteries and problems posed or exhibited by humans.

Remember the joke about the monkeys and the typewriters, where with enough monkeys randomly typing, and time, one of them would write a brilliant novel? Just think of what could happen with intelligent robots that can collaborate and collectively solve great mysteries and problems such as: a) understanding the beginning of the universe where quantum theory apparently dominates; b) how the human race might maintain or improve our quality of life, even as the climate continually changes, and we continue to deplete the Earth's resources; c) how best to at least combat and mitigate what is becoming very rampant terrorism plaguing the civilized World!; and d) retard the human trend to invent and adopt technologies deleterious to human survival, witness nuclear weapons and automobiles, for example. The robots are more likely to succeed at this premise's endeavors to the extent they work together openly and without, e.g. greed, ignorance, jealousy and rancor, in contrast to what human beings are wont to do. But how do we get robots to *behave that way*, e.g. with a moralistic conscience and with empathy?

Premise 3. Robots eventually take over completely, similar to what happened in 'The Planet of the Apes' movie [7].

This may be the more far-fetched and rather negative premise, but it resonates somewhat in contemplating if/when we develop a set of processes (without sufficient safeguards) in evolving a strong, but hopefully not dominant, robotic culture.

2.2 Who?

Consider that humans are getting new body parts, e.g., hip, knee and shoulder replacements, artificial limbs, and vital organ substitutions, e.g., the heart, liver, lungs, (critical portions of) the brain, etc. Maybe by leveraging those accomplishments with continued improvements in medicine to prolong life over a greater period of time, some human beings will become a lot closer to almost turning into robots. Such an evolution may turn selected members of our species into pseudo-robots that could live essentially forever, depending on the reliability, maintainability, and longevity of the body parts, or at least as long as Methuselah!

This gradual body part replacement scenario may greatly ease the angst about robots doing all the jobs (of Premise 1) while humans atrophy with nothing to do. Also, humanity could become more accepting of the idea of robots, if/when religious beliefs mature further, possibly creating momentum to encourage societies to become more atheistic. Other mindsights may gain traction, as well, such as having fewer babies to help the overpopulation problem, and demanding fewer material goods, both of which would slow the depletion of the Earth's resources and mitigate what is likely to be a continual decrease in our quality of life.

2.3 When?

It seems rather obvious that none of the premises posed, at least so far, have much chance of being realized within mere decades. And it is impossible to accurately predict when any of the premises might be realized. Even if we are fortunate enough to suggest some feasible ways forward, we should not expect to pre-specify outcomes or anticipate any timelines for progress. This is the nature of any truly complex system. One can try to influence the system but should be content to observe what happens and patiently await trends of the system's evolution before bravely intervening again.

2.4 Why?

Regarding Premise 1, we already know of debate among some economists asking:

a. What would this do to economies and people?
b. How would people survive?
c. Would people have to earn a living/make money if all labor, food production, teaching, etc., is done essentially for free? If so, how?
d. Would we end up with a largely 'flat society' of plenty, or something much more sinister (see e and h below)?
e. Will we become an oligarchy with only a few 'families' owning everything?

This premise is a very complex thing to fathom.

Not to mention what would happen to marriage and human reproduction if some robots became almost indistinguishable from humans and could have sex.

f. Would many people still choose to have a human spouse over the perfect robot, made or used as a substitute for an imperfect human partner?
g. But what if some people become more robotic with advances in medicine so that they transform into superior beings, and what would they do then?
h. What if they or the robots surpass humans and ultimately take over?

2.5 Where?

We might consider establishing an experimental environment for testing the idea of robots being pervasively integrated into our societies where the laboratory is made extraterrestrial, either on the moon, another planet like Mars, or even within an artificially created earthlike structure which orbits the Earth. One would have to be assured, at least initially, that the robots could not escape this environment and come to Earth before we have developed these premises sufficiently to assure our safety.

Yampolskiy devotes a chapter in his excellent book [8, pp. 145–165] to the AI 'confinement' problem because of its grave dangers, a theme he is quite convincing in espousing, e.g. '… (AGI) research should be considered unethical.' [8, p. 139]

2.6 How?

Considering Premise 1, in one case governments could make the robots, and they would be essentially free. Various questions arise.

a. Then would we just abolish money and have anything we want?
b. What would that scenario look like?

In another case, a few people might own the robot companies privately.

c. If no one works, then who buys the robots?
d. Or how would they be allotted and to whom?

Under any of the premises, we should consider and think about what could go wrong. People hungry for power might be able to corrupt the system so that they could control the robots, and thereby everyone else as well.

e. What kind of safeguards could be put in place?
f. In developing a robotic mind, is it even possible to be assured robotic minds will have only benevolent motives and will not turn against the human race?
g. Even given that, how do we know what to expect robotic behaviors to be?

We would need a pretty comprehensive set of failsafe mechanisms.

Optimistically speaking there are continual technological breakthroughs which help researchers explore how the mind works. [9] With such concerted efforts over long periods of time (see **When** above) it might well be possible to develop robotic minds to be not only smart and creative but also benevolent.

3 PROPOSITIONS

Consider the foregoing as background, primarily. We now come to the main section of this paper where the five propositions are discussed. Liberal use of quotations (using '…'s) from other authors is employed to show various points of view. This author will agree/disagree to certain extents and provide his overall assessments.

3.1 Proposition 1: Could robots become human or even surpass humans?

Will AI efforts in robotics eventually achieve a state where robots possess human-like capabilities of cognition, emotion, learning, etc.? If so, then it may become inevitable that robots would greatly accelerate their capabilities with their own self-learning while leaving humans behind. So much so that there would be grave dangers that the human race will become marginalized, at best, or annihilated, at worst.

- '[I] warn you that artificial intelligence could drive mankind into extinction, and [I'll] explain how that catastrophic outcome is not just possible, but likely if we do not begin preparing very carefully *now*. … consider this [an] invitation to join the most important conversation [we] can have.' [2, p. 16]
- '… AI could [achieve and surpass] human intelligence through the process of recursive self-improvement powers including self-replication, swarming a problem with many versions of itself, super high-speed calculations, running 24/7, mimicking friendliness, playing dead and more. We've proposed that an [ASI] won't be satisfied with remaining

isolated; its drives and intelligence would ... put our existence at risk.' [2, p. 70]
- '[Alan] Turing's wartime assistant, mathematician I. J. Good, suggested that the last invention human beings need to create is the first ultra-intelligent machine. ... Don't panic, though: I can't see that we're any closer to achieving it than we were 50 years ago.' [10, p. 87]
- '... the first ultraintelligent machine is the last invention that man need ever make, provided that the machine is docile enough to tell us how to keep it under control ...' [2, p. 105]
- '[but] ... a real artificial intelligence would be smart enough to not reveal itself.' [10, p. 90]
- '... the human brain actually uses many of the same computation techniques as computers. But ... it's not clear if computers will think as we define it, ... Therefore, some scholars say, artificial intelligence equivalent to human intelligence is impossible.' [2, p. 45]
- 'Our survival, if it is possible, may depend on, among other things, developing AGI with something akin to consciousness and human understanding, even friendliness, built in.' [2, p. 46]
- 'In contrast with our intellect, computers double their performance every eighteen months. So the danger is real that they could develop intelligence and take over the world.– Stephen Hawking, physicist.' [2, p. 148]

Despite Hawking's genius and being one of this author's all-time heroes, there's something missing here! How can a machine become 'intelligent' by just continuing the 'brute force' approach of Moore's Law?! Machines can't evolve as human's have done unless they somehow get intelligence (that special thing that must be there to achieve the ultimate change to ASI). This roadblock seems analogous to Turing's halting problem and Gödel's incompleteness theorem. There are true things within our civilization that we (in our civilization) cannot *prove*, e.g. the statements that machines/robots can – or can never – become intelligent!

Yampolskiy [8] elegantly warns of the ASI dangers. He feels it is quite possibly achievable and has suggested various methods of protecting against it.
- '... Unfortunately, the majority of AI books on the market today talk only about what AI systems will be able to do for us, not to us. I think that this book, which in scientific terms addresses the potential dangers of AI and what we can do about such dangers, is extremely beneficial to the reduction of risk posed by artificial general intelligence (AGI).' [8, p. 184]
- 'Kurzweil isn't concerned about roadblocks to AGI since his preferred route is to reverse engineer the brain. He believes there's nothing about brains, and even consciousness, that cannot be computed. [The theory of computation disputes this though!] In fact, every expert I've spoken with believes that intelligence is computable.' [2, p. 163]
- '... an intelligence explosion may be unavoidable once almost any AGI system is achieved. When any system becomes self-aware and self-improving [but could it?!], its basic drives, ... virtually guarantee that it will seek to improve itself again and again.' [2, p. 163]
- 'If AI researchers do eventually manage to make the leap to AGI, ... the result will [not "just"] be a machine that simply matches human-level intelligence. ... such a system would eventually ... focus its efforts on improving its own design, rewriting its software, or perhaps using evolutionary programming techniques to create, test and optimize enhancements to its design. ... an iterative process of "recursive improvement." ... the ultimate result would be an "intelligence explosion" – quite possibly culminating in a machine thousands or even millions of times smarter than any human being. As Hawking and his collaborators put it, it "would be the biggest event in human history."' [11, pp. 232–233]
- '... the invention of [ASI] – [may be] ultimately prove[en] [There is no way of proving this or the alternative.] impossible or will be achieved only in the very remote future. A

number of top researchers with expertise in brain science have expressed this view. Noam Chomsky … says we're "eons away" from building human-level machine intelligence, and that the Singularity [ASI] is "science fiction." Harvard psychologist Steven Pinker agrees, saying, "There is not the slightest reason to believe in a coming singularity. The fact that you can visualize a future in your imagination is not evidence that it is likely or even possible."' [11, pp. 236–237]

This author sides with Chomsky's and Pinker's expert opinions, though not declaring impossibility, but believing that any possibility for this sort of outcome is many eons away, depending mainly on the further evolution of humans and/or robots. A good sense of this can be acquired by reading Pinker [12] and Dawkins [13], for example.

- '… the problem of AGI is simply too difficult for humans, no matter how long we chip away at it. … we may not possess minds that can understand our own minds. … it might require intelligence greater than our own to fathom our intelligence in full.' [2, pp. 163–164]
- '… we will never achieve AGI … because of the problem of creating human-level intelligence will turn out to be too hard. …' [2, p. 189]

This author tends to believe this quotation, and other similar opinions, so his answer to Proposition 1 is 'No.' The reader is invited to consider the above quotations, delve into further remarks in the cited references, and make up their own mind.

Deterministic software can result in unpredictable behaviors that are unforeseen, i.e. emergent, and even surprising. But that's not the same as creating an AGI being.

Despite remarkable advances in voice recognition and natural language processing, languages are still imprecise and open to interpretation; we often cannot agree on definitions but we can exchange our perceptions of the underlying reality (which no one can truly grasp by themselves) to understand each other better and get a better idea of reality collectively. And partly because of this we cannot instruct a robot on how to become 'intelligent' either!

Maybe our best defense against the dangers of ASI would be in improving our understanding of why even AGI (and thereby ASI) can 'never' be achieved, at least not faster than the prospect of improved humans through further eons of evolution!

3.2 Proposition 2: What if robots take over all the jobs?

Maybe robots will never surpass us or become truly human but what if they become so adept and pervasive that they eventually take over *all* our jobs!?

Even at the accelerated rate robotic applications are heading in this direction, it would likely take a very much longer for us to lose all our jobs. Not that this would be so bad for many who crave more leisure time.

Figure 1 pictorially describes what could be the ultimate reality where robots do all the jobs including the fabrication/maintenance/repair/disposition of robots. Robots perform labor for everyone including the owners of the robots and the government. The owners may include government personnel, and if there are no taxes, the owners would support government directly. The owners possess all wealth, which includes all natural and artificial resources, land and material goods resulting from the means for production and jobs. Robots utilize resources which belong to the few owners of all the robots. Robots provide military protection to everyone, including the government which leads that capability. Of course,

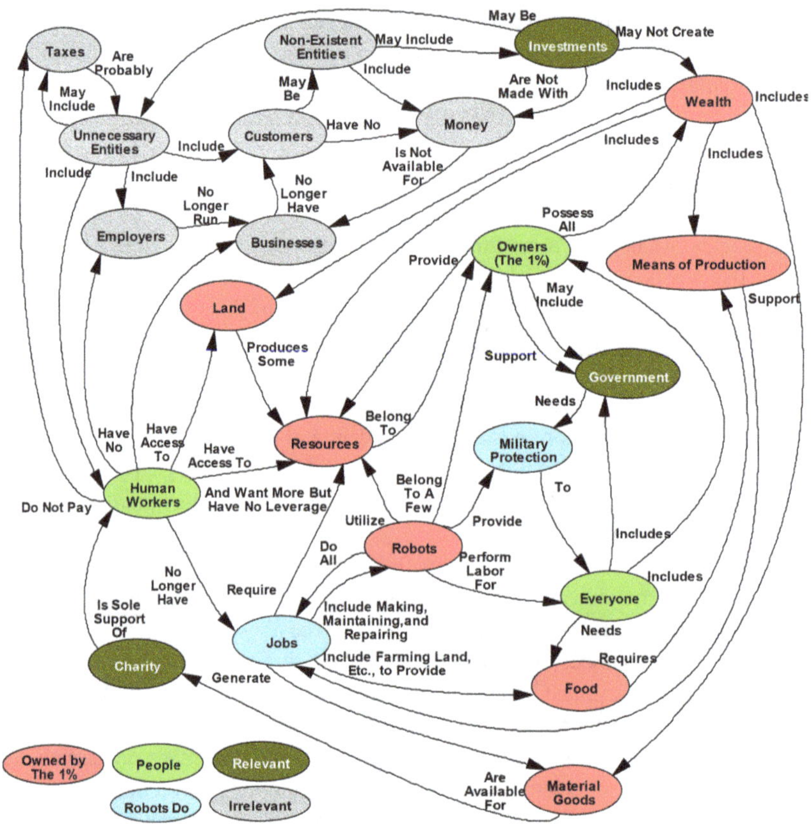

Figure 1: Ultimate reality when robots perform all jobs.

everyone eats food which becomes available through jobs the robots perform using all means for production. Thus, the previously viable human workers no longer have jobs or access to resources. Although they no longer pay taxes and have no employers, these humans do still have access to businesses and land. Humans, as former workers, are solely supported by charity which is derived from material goods and services produced by robots now doing all their jobs. Businesses no longer have customers, who in turn have no money, so businesses cannot prosper and employers cannot run them. Since money does not exist, investments might be made in other ways but still may not create wealth. Unnecessary entities include employers' customers as well as human workers, although customers may also be nonexistent. Taxes are also probably unnecessary.

- 'People don't seem to mind when robots clean their floors, build their cars, administer their medication, or perform their surgery. These examples, however, are all agency-related examples of thinking and doing; there seems to be something very different about a robot that can sense and feel. In robots we seem to be hesitant to combine agency with experience, to make a human machine.' [1, p. 73]
- 'I am profoundly grateful to parents for raising the generation who will be programming the robot to look after me in my decrepitude. ...' [10, p. 204]

Personal healthcare is one area where robots may take longer to replace human caregivers because of the ability of humans to show genuine empathy and emotional support to patients, particularly those with dementia, Alzheimer's disease, etc.

Ultimately, there would be a very small minority of humans that literally own everything, including all the robots. These extremely elite owners would be expected to subsidize the rest of humanity to ensure others' existence at some reasonable levels, or even peoples' survival – or not! What is to prevent these owners from remaining selfish and evilly shirking their moral duty?!

3.3 Proposition 3: Will humans essentially turn into pseudo-robots?

Considering the [im]possibility of robots becoming human or more, as discussed under Proposition 1, what about humans becoming more robot-like? Wouldn't that proposition a lot easier to achieve?

The key to this might be the extent to which medicine advances mimicking portions of the human brain, perhaps sub-human in capability but good enough for robotic behavior. The more challenging aspects of the brain, that just cannot be copied, as yet, would remain human. Voila!, a pseudo-robot (or cyborg, for cybernetic organism [14]) that is still somewhat (mostly?) human.

- 'What exactly will constitute a "robot" when humans augment their bodies and brains with intelligent prosthetics and implants?' [2, p. 21] [One concern: how would efficient power be provided to the artificial parts while the metabolism and muscles of the human portion shrink?]
- '... detrimental effects come from computers *outside* our bodies. Yet Kurzweil proposes only good things will come of computers *inside* our bodies. I think it's implausible to expect that hundreds of thousands of years of evolution will turn on a dime in thirty years, and that we can be reprogrammed ...' [2, p. 147]
- 'Homo sapiens are not known to be particularly harmless when in contact with one another, other animals, or the environment. Who is convinced that humans outfitted with brain augmentation will turn out to be friendlier and more benevolent than machine[s] [with ASI]? ...' [2, pp. 156–157]

It seems quite doubtful that medical and other researchers will ever be able to produce even a partially synthetic brain that guarantees benevolence. Indeed, how does one define such agreeable, caring, and harm-avoiding behavior when any set of postulated 'laws' would likely be at least somewhat ambiguous and not 'air tight?'

3.4 Proposition 4: Will Big Data Authoritarians Psychologically Control Us?

Someday, someone quite autocratic might be able to capture power through a populist vote of the disenchanted who want change. S/he would probably be adept at spreading falsehoods without offering facts, and over a rather short period of time bolster support among her/his populace and hoodwink others. From there it's a relatively easy step to leverage Big Data in achieving '1984' – like control of public thought. [15] In 2012 former US Supreme Court Justice David Souter gave us a somber warning. [16] Here are excerpts.

- ' "an ignorant people can never remain a free people." Democracy cannot survive too much ignorance ... [Benjamin] Franklin was asked by someone ... what kind of government the constitution would give us if it was adopted. Franklin's famous answer was "a republic, if you can keep it." You can't keep it in ignorance. ... What I worry about is that when problems are not addressed, people will not know who is responsible. And when the problems get bad enough, ... some one person will come forward and say, "Give me total power and I will solve this problem."'

- 'Privacy includes our right to keep a domain around us, which includes ... our body, home, property, thoughts, feelings, associations, secrets and identity. The right to privacy gives us the ability to choose which parts in this domain can be accessed by others, and to control the extent, manner and timing of the use of those parts we choose to disclose.' [10, pp. 236–237]
- '... privacy [is divided] into three equally [important] parts: 1) Secrecy—our ability to keep our opinions known only to those we intend to receive them, ... 2) Anonymity – secrecy about who is sending and receiving an opinion or message, and 3) Autonomy – ability to make our own life decisions free from any force that has violated our secrecy or anonymity. Without privacy, nobody can be fully autonomous or free.' [10, p. 237]

3.5 Proposition 5: Will humans keep losing many more jobs to robots?

Despite the addition of millions of jobs in the US under President Obama and a quite low unemployment rate of about 4.7% (as of 9 March 2017), the percentages of those with just part-time jobs is about 9% (though dropping), and those not even looking for work is roughly 12%–15%. The latter two statistics are significantly impacted by automation and robots taking over manual and some white-collar and/or knowledge-based jobs. It should be recognized that this trend will likely continue.

- '... the first decade of the twenty-first century resulted in the creation of no new jobs. Zero. ... [This] is especially astonishing when you consider that the US economy needs to create roughly a million jobs per year just to keep up with growth in the size of the workforce.' [11, p. xi] [A mitigating factor might develop if we started having significantly fewer babies! [17]]
- '... a great many jobs and tasks are likely to be encapsulated in [Big Data] waiting for the day when a smart machine learning algorithm comes along ... As an example, consider radiologists, ... computers are rapidly getting better at analyzing [very large numbers of very complicated] images. It's quite easy to imagine that someday, in the not too distant future, radiology will be a job performed almost exclusively by machines. ... Indeed, ... employment for many skilled professionals—including lawyers, journalists, scientists, and pharmacists—is already being significantly eroded by [AI]. ... most jobs are, on some level, fundamentally routine and predictable, with relatively few people paid primarily to engage in truly creative work or "blue-sky" thinking.' [11, pp. xv–i]
- '... one of the most fundamental ideas woven into the American ethos—the belief that anyone can get ahead through hard work and perseverance—really has little basis in statistical reality.' [11, p. 47]
- 'The hollowed-out middle of the already polarized job market is likely to expand as robots and self-service technologies eat away at low-wage jobs, while increasingly intelligent algorithms threaten higher-skill occupations. ... occupations amounting to nearly half of US total employment may be vulnerable to automation within roughly the next two decades.

... If, as seems likely, advancing technology continues to drive the [US] and other industrialized countries toward ever higher inequality, then the political influence wielded by the financial elite can only increase.' [11, p. 59]

This will become quite telling and tragic for those who voted for Donald Trump in 2016, thinking he will be able to bring back their old jobs or create new jobs. Their job losses are not due to some group of politicians!

- 'The evaporation of thousands of skilled information technology jobs is likely a precursor for a much more wide-ranging impact on knowledge-based employment. ... software will be hosted in the cloud. ... [and software] will eventually be poised to invade virtually every workplace and swallow up nearly any white-collar job that involves sitting in front of a computer manipulating information.' [11, p. 107]
- '... offshoring is very often a precursor of automation, and the jobs it creates in low-wage nations may prove to be short-lived as technology advances. What's more, advances in artificial intelligence may make it even easier to offshore jobs that can't yet be fully automated.' [11, p. 115]

Here's an idea. Maybe tap the wealthy (highly successful) entrepreneurs to share most of their accumulated and future wealth with the general public to help the lower (and even middle) classes survive economically. This would, in part, be payback for some of the public's earlier tax revenue which fueled many innovators with Federal research grants through DARPA (Defense Applied Research Project Agency), NSF (National Science Foundation), universities, small businesses, etc. [11, pp. 80–81]

- 'For progressives, a guaranteed income may be an easier sell in the current political environment. Despite [the conservative market-based approach] argument to the contrary, many liberals would likely embrace the idea as a method to achieve more social and economic justice. A basic income could effectively ... alleviate poverty and mitigate income inequality. At a stroke of the presidential pen [unlikely to happen under Trump's Republican-controlled Congress, of course], extreme poverty and homelessness in the United States might effectively be eradicated.' [11, pp. 259–261]
- 'I don't see anything especially dystopian in offering some relatively unproductive people a minimal income as an incentive to leave the workforce, as long as the result is more opportunity and higher incomes for those who do want to work hard and advance either situation.' [11, p. 269]
- 'Instead, we ought to transition [toward less labor and more capital] taxation that asks more from those businesses that rely heavily on technology and employ relatively few workers.' [11, pp. 277–278] [Let's also do better in retraining threatened workforces and educating our young.]
- 'The widely held belief that a degree in engineering or computer science guarantees a job is largely a myth. An April 2013 analysis by the Economic Policy Institute found that at colleges in the United States, the number of new graduates with engineering and computer science degrees exceeds the number of graduates who actually find jobs in these fields by 50 percent.' [11, p. 127] [Online education is a possible answer if the current gap in the ability of employers to be assured that online participants have really learned something useful to their businesses can be narrowed. [18]]

- 'Unemployment is going to be a serious problem—but not, surprisingly, because of a lack of jobs. Rather, the skills required to do the available jobs are likely to evolve more quickly than workers can adapt without significant changes to how we train our workforce.' [19, p. 13] [Maybe we need Proposition 2 to become a reality!]

4 CONCLUSION

The propositions posed are important to contemplate but are by no means settled. Many authors familiar with artificial intelligence (AI) and the subject of robots have expressed their judgments, as indicated by many of the quotations provided here. After perusing much of the literature on the subject, this author believes

- Humans will never be able to produce human or super-human intelligent robots. This could only happen through natural evolution over eons; if it does there would definitely be an end to human life as we know it.
- Theoretically robots could take over nearly all the jobs performed by humans if present trends continue indefinitely. This could be the ultimate widening of the income gap between the wealthy 1% and the rest of us.
- A modicum of humans will eventually become pseudo-robotic through artificial body-part replacements as medical knowledge and effective practice in this direction continues to blossom.
- Big data techniques will increasingly be leveraged by dictators, autocrats and/or their proponents and pretenders to stultify learning and factual knowledge among the masses for the purposes of psychological control.
- Job losses among human workers will accelerate in favor of robots. An increasing awareness that our methods of education and retraining require revolutionary transformations is necessary to alleviate these effects.

ACKNOWLEDGMENTS

Beaumont Vance posed the question of Proposition 2 which stimulated the author to develop this paper. Thanks to Wayne Davis for several astute comments, and to Matthew Joordens for his interest in collaborating, and a definition of cyborg.

REFERENCES

[1] Wegner, D.M. & Gray K., *The Mind Club—Who Thinks, What Feels, and Why It Matters*, Viking: New York, 2016. [a largely psychological treatment of how we perceive 'mind']

[2] Barrat, J., *Our Final Invention—Artificial Intelligence and the End of the Human Era*, St. Martins Press: New York, 2013. [a pretty scary treatise!]

[3] *The Oxford American College Dictionary*, Putnam: New York, 2002.

[4] Huang, T., 6 questions that can help journalists find a focus, tell better stories. *Poynter*, 9 May 2011, available at http://www.poynter.org/2011/6-questions-that-can-help-journalists-find-a-focus-tell-better-stories/131491/

[5] Zachman, J.A., The Zachman Framework for Enterprise Architecture, Zachman Institute for Framework Advancement (ZIFA), available at www.zifa.com

[6] Kasser, J. & Mackley, T., Applying systems thinking and aligning it to systems engineering. *18th INCOSE International Symposium*, Utrecht, Holland, 2008.

[7] "Planet of the Apes," Movie, Director: Franklin J. Schaffner, Writers: Michael Wilson (screenplay), Rod Serling (screenplay), Stars: Charlton Heston, Roddy McDowall, Kim Hunter, 3 April 1968, available at http://www.imdb.com/title/tt0063442/

[8] Yampolskiy, R.V., *Artificial Superintelligence—A Futuristic Approach,* CRC Press: Boca Raton, 2016. [very "theory of computation" oriented]

[9] Joseph, A., Lighting the way in brain science. *The Boston Globe*, 12 November 2015, p. C5.

[10] Harkness, T., *Big Data—Does Size Matter?* Bloomsbury Sigma: London, 2016. [good account of what is known about us, individually and collectively]

[11] Ford, M., *The Rise of Robots—Technology and the Threat of a Jobless Future*, Basic Books: New York, 2015. [how computers and automation are taking over even knowledge-based jobs]

[12] Pinker, S., *How the Mind Works*, W. W. Norton & Company: New York, 2009. [advocates the credibility of evolution]

[13] Dawkins, R., *The Selfish Gene*, 30th Anniversary edition, Oxford University Press: Oxford, 2006. [includes a treatment of evolution]

[14] Cyborg, Wikipedia, the free encyclopedia, available at https://en.wikipedia.org/wiki/Cyborg.

[15] Orwell, G., *1984—a novel*, Secker and Warburg: London, 1949.

[16] available at http://crooksandliars.com/2016/10/justice-david-souter-civic-ignorance-how.

[17] White, B.E., Applying complex systems engineering in balancing our earth's population and natural resources. *The 7th International Conference for Systems Engineering of the Israeli Society for Systems Engineering (INCOSE_IL)*, Herzlia, Israel, 4–5 March 2013. [suggests incentives for having fewer babies]

[18] White, B.E. & Gandhi, S.J., The Case for Online College Education—a work in progress, *American Society for Engineering Education (ASEE) Annual Conference*, Atlanta, GA, 23–26 June 2013. [promising alternative to having to go but not affording college]

[19] Kaplan, J., *Humans Need Not Apply—A Guide to Wealth and Work in the Age of Artificial Intelligence*, Yale University Press: New Haven, 2015. [on how many common jobs are threatened by artificial intelligence, and related questions of morality, human rights, and income/wealth inequality]

UNCERTAINTY DYNAMICS IN A MODEL OF ECONOMIC INEQUALITY

M.L. BERTOTTI[1], A.K. CHATTOPADHYAY[2] & G. MODANESE[1]
1 Free University of Bozen-Bolzano, Faculty of Science and Technology, Bolzano, Italy.
2 Aston University, Mathematics, System Analytics Research Institute, Birmingham, UK.

ABSTRACT

In this article, we consider a stylized dynamic model to describe the economics of a population, expressed by a Langevin-type kinetic equation. The dynamics is defined by a combination of terms, one of which represents monetary exchanges between individuals mutually engaged in trade, while the uncertainty in barter (trade exchange) is modeled through additive and multiplicative stochastic terms which necessarily abide dynamical constraints. The model is studied to estimate three meaningful quantities, the inequality Gini index, the social mobility and the total income of the population. In particular, we investigate the time evolving binary correlations between any two of these quantities.

Keywords: *additive and multiplicative noise, economic inequality, income distribution, social mobility.*

1 MOTIVATION AND INTRODUCTORY CONSIDERATIONS

The emergence of inequality in income and wealth distribution has attracted considerable interest in recent years. Books by renowned economists addressing this issue (e.g., [1–3]) are having a widespread diffusion among non-specialists too, with frequent tectonic societal impact generating substantial media highlights. Besides, the urge to identify possible mechanisms for the formation of such collective phenomenon prompted the formulation of related models within an enlarged scientific community (see e.g. [4–9]).

Typically, in these models, society is dealt with as a system composed of a large number of individuals who exchange money through binary and other nonlinear interactions. For example, the model proposed in [8] (see also [10]) by two of the present authors is expressed by a system of nonlinear differential equations describing the evolution in time of the income distribution. Specific trading rules which characterize the behavior of individuals of different income classes together with the existence of a taxation and a welfare system are postulated. Despite the stochasticity element due to the presence of transition probabilities for trade and resulting class change of single individuals, the equations governing the variation in time of the fraction of individuals in each income class are deterministic.

An important concept not included in Ref. [8] is that of uncertainty. In almost all walks of life, the presence of chance is unavoidable, at least to some extent. Incorporating this element to analyze the trade dynamics is then crucial. To keep also uncertainty into consideration, we developed and analyzed in Refs [11] and [12] a stochastic model containing an Ito-type additive and an Ito-type multiplicative noise term, respectively [13]. Exploring the resulting dynamics, we were able to recover, at least in some cases, empirically observed patterns. Pondering on all that led us to the conclusion that a more realistic modelling should involve a combination of additive and multiplicative stochastic perturbation. This is what we are doing in this paper.

We stress that, in order to concentrate on the role of noise in the system, here we omit terms describing taxation and redistribution with welfare which were included in Ref. [8]: in other words, the 'deterministic' component of the present model is simpler (and so was that of the models in [11] and [12]). We limit ourselves to consider the occurrence of a wealth of monetary exchanges between individuals and to this we now add the effect of additive and multiplicative stochastic terms. Another difference with the approach developed in [8, 11, 12] is the following. Whereas the variables in these papers describe the fractions of individuals belonging to different income classes, the variables here are the incomes of the classes. This implies that the random perturbations which we are including here directly affect the income of the classes, sectorially and over the whole economic spectrum.

Our aim is to provide a framework to estimate some important indicators measuring respectively economic inequality and social mobility. In particular, we investigate here the (varying in time) correlations between these two indicators and between each of them and the total income of the population.

The paper is organized as follows. In Section 2, we describe the structure of our model, which is expressed by a Langevin-type kinetic equation and we also propose an algorithm generating a combination of additive and multiplicative noise. The equations are then numerically solved and in Section 3 we report results obtained by taking the average of various quantities over a large number of realizations. Therein, we also compare our results with some real world data [14]. Finally, Section 4 provides some concluding remarks.

2 THE EQUATION STRUCTURE

Consider a population of individuals divided into a finite number n of classes, each one characterized by its average income r_j with $0 < r_1 \leq r_2 \leq \ldots \leq r_n$. Let $x_j(t)$, with $x_j : \mathbf{R} \to [0, +\infty)$ for $1 \leq j \leq n$, denote the fraction at time t of individuals in the j-th class and let $y_j(t) = r_j\, x_j(t)$, with $y_j : \mathbf{R} \to [0, +\infty)$ be the total income of the class j. In previous work, see e.g. Ref. [8], two of us constructed a model for the evolution in time of $x(t) = (x_1(t),\ldots, x_n(t))$, in correspondence to a whole of economic exchanges taking place between pairs of individuals. From specific behavioral assumptions on individuals of different classes expressed through different transition probabilities, we were led to write down a system of ordinary differential equations with $x(t)$ as unknown function, in fact 'deterministic' in the x_j variables. Subsequently, we also included in our models the presence of an additive [11] or a multiplicative [12] noise term. In this paper, we aim to discuss a different model, expressed by a Langevin kinetic equation for which both additive and multiplicative noise terms are present and for which the variable is the income vector $y(t) = (y_1(t), \ldots, y_n(t))$. Specifically, the equations we consider take the form

$$dy_j = A_j(y)dt + B_j(y,\eta^A,\eta^M)\sqrt{dt}, \quad 1 \leq j \leq n, \tag{1}$$

in which the 'deterministic' part

$$A_j(y)dt = \left(\sum_{h,k=1}^{n} A_{hk}^{j} y_h y_k - y_j \sum_{i=1}^{n} y_i \right) dt \tag{2}$$

is obtained through a reformulation of the r.h.s. of the equations in Ref. [8] (we point out, however, that we disregard here, for sake of simplicity, taxation and redistribution terms) and the stochastic part

$$B_j\left(y,\eta^A \eta^M\right)\sqrt{dt} = \left(\omega\eta_j^A + (1-\omega)\eta_j^M\right)\Gamma\sqrt{dt} \tag{3}$$

describes an Ito-type process [13] incorporating additive and multiplicative stochastic noises. In more detail, the coefficients in the homogeneous quadratic part of Aj (y) are given by

$$A_{hk}^i = \frac{r_i}{r_h r_k} C_{hk}^i \tag{4}$$

with $C_{hk}^i \in [0,1]$ for $i,h,k \in \{1,\dots,n\}$ as in the formula (2) in Ref. [8], expressing the probability that an individual of the h-th class will belong to the i-th class after a direct interaction with an individual of the k-th class. In particular, in the expression for $A_j(y)$ in Equation (2) the identity $\sum_{i=1}^n C_{hk}^i = 1$ for any $h,k \in \{1,\dots,n\}$ is used.

As for terms in equation (3), $\eta^A = \left(\eta_1^A,\dots,\eta_n^A\right)$ and $\eta^M = \left(\eta_1^M,\dots,\eta_n^M\right)$ denote an additive and a multiplicative noise vector, respectively, $\omega \in [0,1]$ is randomly chosen at each integration step, and Γ denotes the noise amplitude. As in Ref. [8], we assume here constant population size during the evolution of the system and normalize it to 1: $\sum_{j=1\dots n} x_j(t) = 1$ for all $t \geq 0$. For this to occur, the noise vectors have to satisfy a suitable constraint (also the terms in the deterministic part of the equations have been reconstructed so as to satisfy the required condition):

$$\sum_{j=1}^n \frac{\omega \eta_j^A + (1-\omega)\eta_j^M}{r_j} = 0 \tag{5}$$

In view of this, the numerical algorithm reciprocating the dynamics can be represented as follows: at each step one chooses two vectors $\zeta = (\zeta_1,\dots,\zeta_n)$ and $\xi = (\xi_1,\dots,\xi_n)$ whose components are Gaussian random variables, and then, starting from these, one can define the vectors η^A and η^M by setting

$$\eta_j^A = \zeta_j - \frac{1}{r_j} \frac{\sum_{i=1}^n \frac{\zeta_j}{r_j}}{\sum_{k=1}^n \frac{1}{r_k^2}} \tag{6}$$

$$\eta_j^M = y_j \xi_j - \frac{y_j^2}{r_j} \frac{\sum_{i=1}^n \frac{y_j \xi_j}{r_j}}{\sum_{k=1}^n \frac{y_k^2}{r_k^2}} \tag{7}$$

It is easy to verify that the vectors η^A and η^M as in Equations (6) and (7) satisfy $\sum_{j=1}^n \frac{\eta_j^A}{r_j} = 0$ and $\sum_{j=1}^n \frac{\eta_j^M}{r_j} = 0$.

Finally, the parameter ω is randomly chosen in [0,1] and Γ tunes the noise amplitude.

3 NUMERICAL RESULTS

Our aim is to evaluate the Gini index G and another indicator, M, which quantifies social mobility during the evolution of equations (1). We are interested as well in getting information on the sign of the correlation of G and M, and on the sign of their correlations with the value μ of the total income of the population.

We recall that the index G was proposed by the Italian statistician Corrado Gini a century ago as a measure of inequality or income or wealth. It is defined as a ratio, whose numerator is given by the area between the Lorenz curve of a distribution and the uniform distribution line, while the denominator is given by the area of the region under the uniform distribution line. It takes values in $[0,1]$.

As for the definition of M, introduced in Ref. [15] in a partially different context, we need to recall first a couple of notations it involves. In Refs [8] and [15] (and the same, albeit not explicitly mentioned, holds here in view of the expressions of the coefficients C_{hk}^i which are, as written in Section 2, as in the formula (2) in Ref. [8]) we denoted S to be the amount of money exchanged in each trade, and we introduced, in order to bring heterogeneity into the model, suitable parameters $p_{h,k}$ for $h, k = 1,\ldots, n$, with each $p_{h,k}$ measuring the encounter frequency rate of individuals of the h-th and in the k-th class and expressing the probability that in an encounter between an h-individual and a k-individual, the one who pays is the h-individual. Then, M can be defined as

$$M = \frac{1}{\left(1 - \dfrac{y_1}{r_1} - \dfrac{y_n}{r_n}\right)} \sum_{i=2}^{n-1} \sum_{k=1}^{n} \frac{S}{(r_{i+1} - r_i)} \, p_{k,i} \, \frac{y_k}{r_k} \frac{y_i}{r_i}$$

namely as the collective probability of class advancement of all classes from the 2-th to the $(n-1)$-th one.

For the purpose of computing G and M, we solved numerically the equations (1) and took the average of the quantities of interest out of a large ensemble of stochastic realizations. More specifically, in our simulations, we considered n = 10, $r_1 = 10$ and $r_i - r_{i-1} = 10$ for $2 \leq i \leq n$, Γ equal to 0.001, and, as already pointed out, we chose randomly $\omega \in [0,1]$ at each integration step. We always considered a stationary solution, reached in the long time, of the equations (1) with $\Gamma = 0$, i.e. in the absence of noise as initial condition. In every simulation, we ensembled averaged over 50 realizations with each run spanning 5000 integration steps.

A few samples of the results we obtained are reported in Table 1. Here, nine triplets are displayed with the average values of correlations R_{GM} (Gini and mobility index), $R_{G\mu}$ (Gini

Table 1: Correlations R_{GM} (Gini and mobility index), $R_{G\mu}$ (Gini index and total income) and $R_{M\mu}$ (mobility index and total income) computed in nine cases in which total income μ is not conserved, with noise amplitude $\Gamma = 0.001$. Averages of 50 realizations, each of 5000 integration steps.

$\mu(0)$	R_{GM}	$R_{G\mu}$	$R_{M\mu}$
25	-0.471 ± 0.058	-0.125 ± 0.068	0.901 ± 0.013
25	-0.518 ± 0.045	-0.136 ± 0.055	0.883 ± 0.016
25	-0.448 ± 0.053	-0.110 ± 0.065	0.906 ± 0.014
30	-0.617 ± 0.046	-0.369 ± 0.057	0.938 ± 0.010
30	-0.641 ± 0.041	-0.404 ± 0.055	0.942 ± 0.010
30	-0.642 ± 0.047	-0.442 ± 0.060	0.955 ± 0.010
35	-0.711 ± 0.034	-0.588 ± 0.047	0.979 ± 0.003
35	-0.711 ± 0.041	-0.601 ± 0.030	0.980 ± 0.004
35	-0.696 ± 0.038	-0.580 ± 0.050	0.982 ± 0.003

index and total income) and $R_{M\mu}$ (mobility index and total income). These values are computed using solutions evolving from three initial conditions for which the initial value of μ is equal to 25, 30 and 35, respectively. For any of these initial conditions, three different average results are reported.

As should be evident from the first column of data, the correlation between economic inequality and social mobility represented by R_{GM} is negative for the three mentioned values of μ. Most remarkably, this negative correlation mimics real world situations [16, 17]. Here, one might wonder what is the meaning of these values of μ. To answer that question, we point out that a relation between the total income μ and the Gini index G can be seen to hold true at equilibrium, approximately given by $G = -0.1594 + 0.03712\mu - 0.0006\mu^2$. Hence, the data in the table can also be thought of as relative to three cases with G close to 0. 39, 0.41 and 0. 40, respectively. A graphical, quite expressive illustration of the sort of values of R_{GM} we obtained for G (randomly chosen) in the range $0.35 < G < 0.41$ is provided in Figure 1. Each dot therein represents the average over 40 stochastic realizations with 5000 integration steps.

The central data column in Table 1 provides particular values of the negative correlation between G and μ. The corresponding right column shows that there is a strong positive correlation $R_{M\mu}$ between mobility and total income.

Finally, we emphasize that the range of values of G referred to in Figure 1 includes the Gini indices of various countries. These can be found on the web page of the World Bank [14]. For example, the Gini index of the United States whose most recent reported value is relative to the year 2013, is 41.06. The Gini indices (values relative to the year 2012) of the countries where we live, Italy and UK, are 35.16 and 32.57, respectively.

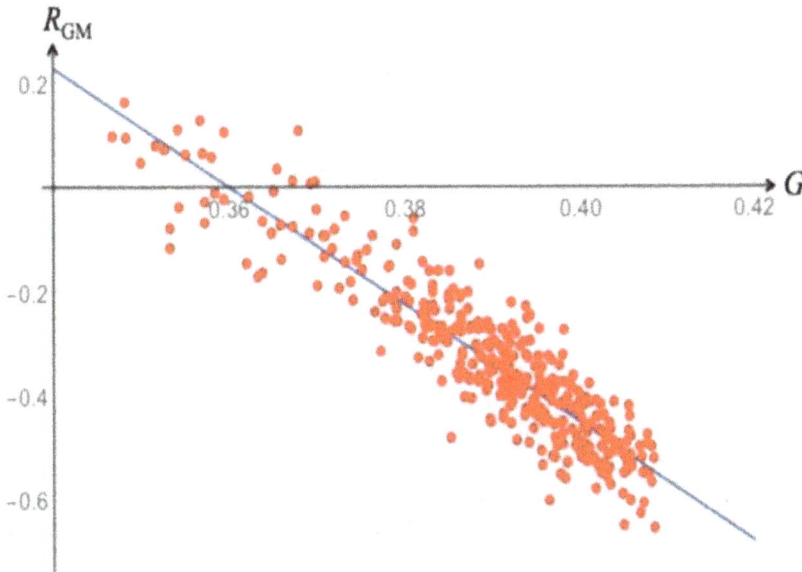

Figure 1: Correlation R_{GM} for different values of G in the range $0.35 < G < 0.41$. Each dot represents the average of 40 realizations of stochastic time series with 5000 integration steps. The equation of the regression line is $R_{GM} = -11.35\,G + 4.085$.

4 CONCLUDING REMARKS

A stylized dynamic model has been proposed and numerically investigated, which allows to estimate economic inequality and mobility for a population. The model describes a range of economic exchanges between population members driven by a combination of additive and multiplicative stochastic noises, resembling uncertainty in the trade situation. The resulting Langevin-type kinetic equation is represented by a minimalist combination of 'deterministic' and stochastic components. The deterministic part assumes phenomenologically supported rules [8] (supposed to be the same for individuals in the same income class) of economic exchanges between pairs of individuals. We emphasize that they do not include effects such as taxation and redistribution, which were studied in other papers [8, 10]. While admittedly our structure does not correspond to the complex nebular real-world interactive description, some interesting results are found in connection with the sign of the indicators of inequality and mobility and especially in connection with their correlations.

REFERENCES

[1] Stiglitz, J.E., *The price of inequality: how today's divided society endangers our future*, W.W. Norton & Company, New York, 2012.

[2] Atkinson, A.B., *Inequality: what can be done?* Harvard University Press, Cambridge, 2015.

[3] Milanovic, B., *Global inequality. A new approach for the age of globalization*, Harvard University Press, Cambridge, 2016.

[4] Angle, J., The inequality process as a wealth maximising process. *Physica A*, **367**, pp. 388–414, 2006.
https://doi.org/10.1016/j.physa.2005.11.017

[5] Matthes, D. & Toscani, G., On steady distributions of kinetic models of conservative economies. *Journal of Statistical Physics*, **130**, pp. 1087–1117, 2008.
https://doi.org/10.1007/s10955-007-9462-2

[6] Chakrabarti, A. & Chakraborti, B.K., Microeconomics of the ideal gas like market models. *Physica A*, **388**, pp. 4151–4158, 2009.
https://doi.org/10.1016/j.physa.2009.06.038

[7] Yakovenko V.M., Econophysics, statistical mechanics approach to, *Encyclopedia of Complexity and System Science*, edited by R. A. Meyers, Springer, 2009.

[8] Bertotti, M.L. & Modanese, G., Microscopic models for welfare measures addressing a reduction of economic inequality. *Complexity*, **21**(6), pp. 89–98, 2016.
https://doi.org/10.1002/cplx.21669

[9] Chattopadhyay, A.K., Krishna Kumar, T. & Mallick, S.K., Poverty index with time-varying consumption and income distributions. *Physical Review E*, **95**, 032109, 2017.
https://doi.org/10.1103/physreve.95.032109

[10] Bertotti, M.L. & Modanese, G., Exchange models for the emergence of income distribution and economic inequality. *International Journal of Design & Nature and Ecodynamics*, **11**(4), pp. 620–627, 2016.
https://doi.org/10.2495/dne-v11-n4-620-627

[11] Bertotti, M.L., Chattopadhyay, A.K. & Modanese, G., Stochastic effects in a discretized kinetic model of economic exchange. *Physica A*, **471**, pp. 724–732, 2017.
https://doi.org/10.1016/j.physa.2016.12.072

[12] Bertotti, M.L., Chattopadhyay, A.K. & Modanese, G., Economic inequality and mobility for stochastic models with multiplicative noise, preprint, arXiv:1702.08391, 2017.

[13] Risken H., *The Fokker-Planck equation*, Springer Verlag, Berlin, 1984.

[14] available at: http://data.worldbank.org/indicator/SLPOV.GINL

[15] Bertotti, M.L. & Modanese, G., Economic inequality and mobility in kinetic models for social sciences. *The European Physical Journal Special Topics*, **225**, pp. 1945–1958, 2016.
https://doi.org/10.1140/epjst/e2015-50117-8

[16] Andrews, D. & Leigh. A., More inequality, less social mobility. *Applied Economics Letters*, **16**, pp. 1489–1492, 2009.
https://doi.org/10.1080/13504850701720197

[17] Corak, M., Income inequality, equality of opportunity, and intergenerational mobility. *Journal of Economic Perspectives*, **27**, pp. 79–102, 2013.
https://doi.org/10.1257/jep.27.3.79

INTEGRATING IN-VEHICLE, VEHICLE-TO-VEHICLE, AND INTELLIGENT ROADWAY SYSTEMS

C. WARREN AXELROD
US Cyber Consequences Unit, USA.

ABSTRACT

With the inexorable push toward autonomous road vehicles by companies such as Tesla Motors and Google, there is an urgent need to make roadways 'smart' and to connect vehicles' computer systems to one another and to their surroundings, infrastructure, and ecosystem. The effective integration of these systems is a major challenge for companies and government agencies. We examine the state of current and evolving systems and communications with respect to in-vehicle (IV), vehicle-to-vehicle (V2V), vehicle-to-surroundings (V2S), vehicle-to-infrastructure (V2I), and vehicle-to-ecosystem (V2E). We define the term 'surroundings' as the immediate vicinity of a vehicle; 'infrastructure' as the local area, such as a municipality or nearby countryside; and 'ecosystem' as distant facilities, such as the Internet, the Cloud, and call centers. We postulate that, for self-driving road vehicles to be fully effective, these efforts must progress together, and selected approaches must be standardized, preferably globally, so that diverse systems can be readily integrated into systems-of-systems. Failure to make such advances across the board will hamper the design, development, and deployment of the many safety-critical systems in need of integration. We suggest how to introduce such technologies, taking into consideration the latest advances and the cost and ease of implementation and support.

Keywords: adaptive, autonomous, complex, complicated, driverless, in-vehicle, self-driving, self-organizing, systems-of-systems, vehicle and traffic control systems, vehicle-to-infrastructure, vehicle-to-vehicle.

1 INTRODUCTION

Ground vehicle manufacturers and software builders are pushing the capabilities of in-vehicle (IV) autonomous systems to their limits in anticipation of enormous profits to be derived from self-driving vehicles [1]. However, system malfunctions and failures have already led to some serious and fatal accidents. We cannot expect to achieve death-proof motoring, as expressed by Volvo in [2], with our current complicated and unintelligent infrastructure [3] and with any residual human element. Human driver involvement cannot be eliminated unless virtually all vehicles are autonomous, can communicate with one another, operate on infrastructures that communicate with vehicles, and are managed by overarching integrated vehicle and traffic control systems. Until these goals are accomplished, the need for human intervention will continue.

The evolution of IV, driver-assist and vehicle-control technologies is fast and furious. Software companies, such as Google, and vehicle manufacturers, notably Tesla Motors, General Motors, Audi, and Volvo, are expanding frontiers with respect to IV technologies. However, until recently, auto manufacturers and government agencies have generally ignored vehicle-to-vehicle (V2V) and vehicle-to-infrastructure (V2I) communications. There are a few exceptions to this assertion, such as a large-scale test of V2V and V2I communications for more than 3,000 cars in Ann Arbor, Michigan, and the introduction of external vehicle communications in some 2017 Audi and Cadillac automobiles, as described in [4]. The paucity of such tests is largely because no single entity establishes, maintains, and enforces interoperability standards among in-vehicle and ex-vehicle systems, which are being developed by vehicle and software manufacturers, and traffic control systems that are being developed by third parties and implemented by government agencies, although some agencies have

© 2018 WIT Press, www.witpress.com
DOI: 10.2495/DNE-V13-N1-23-38

made attempts to come up with standards [5, 6]. There is a critical need for collaboration and coordination among all involved in developing related systems to avoid inconsistencies and incompatibilities.

While IV detection and response technologies are effective over a wide range of traffic situations and under many road conditions, they are limited in scope because, first, vehicles incorporating these technologies currently represent only a small fraction of total vehicles in service and, second, present-day driver-assist and vehicle-control systems are necessarily conservative to allow for deficiencies in the infrastructure and in the preponderance of traditional non-autonomous vehicles. We must consider what will be the outcome when autonomous-vehicle technologies become pervasive and the likelihood of errors, malfunctions, and failures of these highly complex systems increases. Such luminaries as Elon Musk, the founder of Tesla Motors, and such government agencies as the U.S. Government Accountability Office [7], believe that self-driving technologies will greatly reduce the number and seriousness of road accidents, as do some government agencies. While this is clearly a good goal for the automotive industry, there remain many liability, ethical, and moral issues that still need to be resolved.

With no across-the-board standards and significantly different commitments by participants, bringing such diverse systems and constituencies together will be a technical, structural, and organizational challenge. We will discuss the current state of affairs and the likely evolution over short and longer terms, and suggest initiatives that would achieve safe and secure, unified, and integrated autonomous-vehicle systems within a reasonable timeframe.

2 WHERE WE ARE TODAY

We are seeing the rapid development of IV systems. Some are driver-assist systems that monitor traffic and roadway infrastructures but rely on drivers to take appropriate action, as described in [8]. For example, a recent television commercial for Audi cars shows a prototype vehicle operating without a driver, but then insists that drivers are required. Other systems, such as Tesla's Autopilot system, can operate vehicles autonomously, but require driver intervention from time to time. Following a highly publicized fatal accident on May 7, 2016, Tesla updated its Autopilot system to prevent the same occurrence in the future. Tesla was cleared of responsibility for the crash, as described in [9]. Yet others, such as Google, are developing completely autonomous vehicles in which there are no driver controls for steering, accelerating, braking, etc. Some vehicle-to-vehicle systems have already been designed and developed and are in test mode, but there needs to be greater standardization and coordination of these efforts before they can be generally accepted. Furthermore, some individual or some group must decide who will pay for vehicle-to-vehicle communications, which is an issue raised in [10]. There have been several experiments with intelligent roadways on small stretches of road [11, 12], but these are in their early stages and will likely not be widely deployed for a decade or two.

2.1 'Smart' roadways vs. intelligent traffic systems

It is important to differentiate between traditional ITSs (intelligent traffic systems) and newer so-called smart roadways. For one, while ITSs monitor and control traffic, in general, they do not currently communicate with IV systems. If there is an incident and/or congestion, human drivers are made aware through notifications on roadside message boards or are directed to tune into local radio stations for announcements rather than from IV alerts. Drivers can rely somewhat on their IV, dedicated, or smart-phone navigation systems to recommend alternative routes to avoid delays.

3 DEFINITIONS AND ACRONYMS

The terms V2V (vehicle-to-vehicle), V2I (vehicle-to-infrastructure) and V2X (vehicle-to-everything) are in common use [13], but they do not fully reflect the scope and complexity of real-world interactions. Furthermore, they do not provide the granularity needed to describe the full range of possibilities. Consequently, we subdivide and rearrange the V2I and V2X categories into more detailed groups, adding vehicle-to-surroundings (V2S) and vehicle-to-ecosystem (V2E) to reflect current and prospective local, remote, and global interactions. We also added the classifications of surroundings-to-infrastructure (S2I) and infrastructure-to-ecosystem (I2E) to include communications not with vehicles directly. The above classifications are defined and described in Table 1.

Table 1: In-vehicle, vehicle-to-other and other-to-other communications.

Term	Meaning	Description	Comments
IV	In-Vehicle	One-way communications from in-vehic le sensors and external sources.	Sources of information may be installed in the vehicle, or come from external sources.
V2V	Vehicle-to-Vehicle	Two-way communications between vehicles within a limited geographical area.	Vehicle-to-vehicle interactions cover a limited range, within, say, 300 yards.
V2S	Vehicle-to-Surroundings	One-way communication to vehicles from local sources. Will likely evolve into two-way communications.	Category is within the usual definition of V2I. Communications via wireless signals or image recognition.
V2I	Vehicle-to-Infrastructure	One-way communications to vehicles from sources within several miles. Will likely evolve into two-way communications.	Category included within the usual definition of V2I. Includes radio communications reporting incident(s) taking place beyond immediate surroundings but within several miles of a vehicle's location.
V2E	Vehicle-to-Ecosystem	One-way communications to vehicles from external sources, such as GPS. Two-way communications between vehicles and external services.	Category included in the usual definition of V2I. Covers communications with distant sources via wireless, broadband, satellite.
S2I	Surroundings-to-Infrastructure	One-way communications, e.g. weather in area. Two-way communications e.g. where infrastructure informs surroundings of activities beyond the immediate area.	Category has not received much attention to date. However, it is likely that the surroundings will inform the infrastructure of events in greater detail than vehicles might provide.
I2E	Infrastructure-to-Ecosystem	One-way communications, such as local weather. Two-way communications, e.g. infrastructure informs navigation systems of intended actions.	Category does not appear to have been addressed in the literature but municipal traffic control systems will likely communicate to and from the ecosystem.

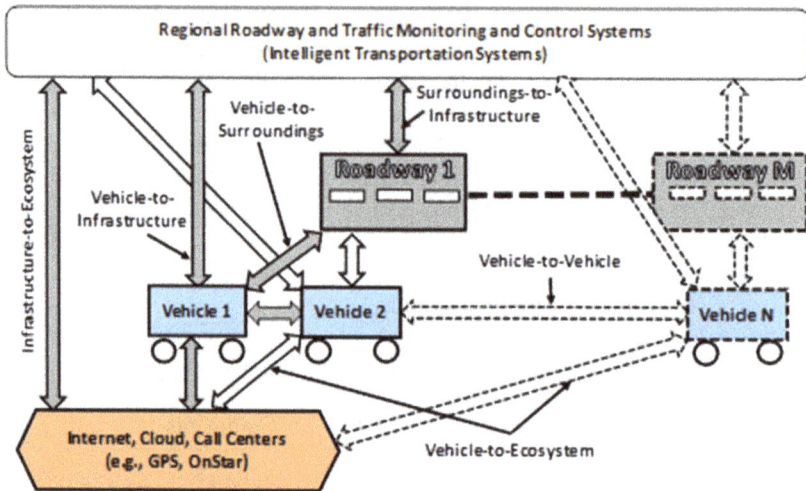

Figure 1: Communications between levels.

These categories are illustrated in Fig. 1. The diagram shows both current and future interactions. The shaded arrows to the left of the diagram show interactions between Vehicle 1 and its surroundings, as well as with the infrastructure and ultimately the broad ecosystem. As mentioned in Table 1, some of the interactions are currently one-way but will likely evolve into two-way communications. As the systems at one level expand, interconnect, and interoperate with systems at other levels, we will see systems-of-systems developing and gradually taking control across the board. This will place significant stress on current roadway operational conditions, especially as the pressure for advancing technologies varies greatly from one level to the next. We will discuss the impact of different rates of adoption and adaptation, the consequences of failing to coordinate projects, and the critical path to achieving fully automated ground transportation systems.

Figure 1 shows various levels within the overall structure and how they might communicate among themselves. The scope of systems at each level will depend greatly on the determination of where specific functions should reside. For example, IV or Internet-based navigation systems may show the speed limits along sections of roadways being travelled. Speed-limit data for Internet-based navigation will have been entered through data entry from observation, whereas IV image recognition will eventually read actual road signs and determine the appropriate speed, both types of system indicating whether the speed limit is being adhered to and, if not, alerting drivers to slow down. The former method depends for accuracy on timeliness and accuracy of data entered, whereas the latter depends on the road signs being visible and not being covered with foliage or snow, or knocked down. There should be decision rules across systems for each situation and the means of resolving any conflicts. Information about current speed and hard acceleration and braking may also be transmitted from IV systems to the ecosystem for insurance purposes, as in the case of the OnStar Smart-Driver system described below, or by law enforcement.

We now examine each of the above categories in greater detail.

3.1 In-vehicle systems

This is the field involving greatest innovation and shortest time-to-market. There are numerous companies pursuing the development of IV systems to provide driver assistance and direct vehicle control, and several government agencies are supporting research and working on standards. On the other hand, there is not much collaboration among organizations within the private sector and between them and the government sector, leading to disparities among systems that will hinder the acceptance and implementation of these technologies across the board. IV sensing and reporting systems could be proprietary if communications protocols are standardized and systems can interoperate with each other and with their environment. An excellent description of IV systems and communications and issues relating to coordination can be found in [5].

3.2 Vehicle-to-vehicle systems

When it comes to V2V systems, we can determine what information they should collect, and what they should do with it. One question that arises is whether data for traffic jams, black ice, and accidents, for example should rely on IV or V2V or V2S systems, or on a combination of, say, V2V and V2S systems. Currently, because of the deployment of notice boards, the main source is probably V2S systems, but navigation systems also announce delays along a predetermined route, presumably obtained from a representative group of vehicles using the same navigation systems, such as Google Maps with the Waze app, which report the speeds of vehicles and use those data to determine that there are delays. They then suggest alternative routes.

3.3 Vehicle-to surroundings systems

Currently, most information is communicated to drivers from the immediate vicinity via human vision. Such information includes recognition of fixed road signs, such as stop, yield, direction signs, variable-message signs providing warnings (such as delays ahead), alerts (such as AMBER alerts) and information, flashing lights of police and road service vehicles, etc. A more complete list for variable-message boards, from Wikipedia, is as follows:

- Road works
- Incidents affecting normal traffic flow
- Non-recurring congestion
- Closure of an entire road
- Exit ramp closures
- Debris on roadway
- Vehicle fires
- Short-term construction
- Pavement failure alerts
- AMBER, Silver, and Blue alerts
- Travel times
- Variable speed limits
- Car park occupancy

As IV systems evolve, they will likely have increased capability to recognize signs through shape determination and word recognition. In fact, a recent video from Volvo, touting their technologies, shows roadside speed-limit sign recognition, reading, and alerting in action [14]. Such systems could potentially warn drivers to slow down for an upcoming stop sign, for example, and apply the brakes, if necessary.

At the same time, we might see roadways becoming more communicative with, for example, signs transmitting data regarding their functions. This could help address the problems of signs being knocked down or covered with tree branches, and the like. It would also reduce the need for IV systems to interpret signs from visual information. There are already prototypes of communications between vehicles and traffic signals as described further.

3.4 Vehicle-to-infrastructure systems

We are seeing some initial forays into V2I systems. It should be noted that these communications will usually be between vehicles and central control systems, such as those controlling traffic lights for a city, as described in [4] for the collaboration between Audi and Traffic Technology Services (TTS). TTS obtains data from local authorities' systems and analyzes it to predict when specific signals will change.

3.5 Vehicle-to-ecosystem systems

Communications between vehicles and distant services have been operational for some time, as with satellite-based GPS (global positioning systems) and call centers (e.g. OnStar). The services can be one-way as with GPS or two-way as with requests for navigation services. Services such as EZpass, which automates the toll payment process, are real-time one-way, in that IV sensors send details (such as location of toll, time of day) and the central system checks the validity of IV sensors and the state of account balances and the use of the service for billing purposes. In this respect, they are long-term two-way when vehicles' owners are charged for the tolls.

This category includes interchanges between vehicles and the Internet, (e.g. Google Maps), the Cloud, call centers, and the like. The services so provided include navigation services, vehicle status reporting, unlocking of doors, automatic reporting of accidents and informing of first responders, tracking and deactivating stolen vehicles, hands-free telephone service, etc. This area is characterized by rapid additions to services, such as the General Motors OnStar Smart Driver System described further, and the transfer of functions among vehicles, their immediate surroundings, and their proximate infrastructure to the Cloud, as has happened with navigation systems.

3.5.1 Traffic information systems

It is valuable for drivers to be notified of heavy traffic ahead and how to avoid the congestion via alternative routes to shorten travel time and reduce stress, fuel consumption, and the like.

When it comes to delays along the route, there are several means of notifying drivers. Local authorities will post the location of congestion and the potential duration of delays on variable-message boards along the highway and/or they will suggest to drivers, by means of flashing lights on a fixed message board, that they tune into a local radio station to obtain information about delays and the like. Notices are generated based on information from sensors and television cameras along the roadside.

Independent navigation systems, such as Google Maps, have evolved in their means of collecting and disseminating congestion data and alternative routing [15]. Earlier congestion was based upon historical data for roadways, which were obtained for time of day and day-of-week, as well as for holidays and various weather conditions. However, as the use of certain navigation systems has progressed, companies such as Google can obtain specific location and travel-speed information from a growing population of vehicles using their system, determine in real-time where delays are occurring, and feed that information back to drivers. Google's acquisition of Waze has provided additional driver-reported information on traffic congestion and alternative routing.

It is noteworthy that the integration of these systems is taking place IV where the human driver can combine information from sophisticated travel-condition systems, less-advanced (and less useful) notice boards, and direct visual observation. It is to be expected that these systems will improve over time and will likely be fully automated as V2V and V2S communications become more common and either character recognition of roadside displays or transmissions from those displays become available to a central system. Whether Google owns such a system, or it is owned by a combination of Google and local authorities, or some other combination, must be worked out. However, it is likely that Google and/or other companies will own and operate such a system based on funding and on the ability to monetize the data collected.

3.5.2 OnStar regular and smart-driver systems

The OnStar system is proprietary to General Motors (GM), although other vehicle brands have either licensed the OnStar system or have developed their own equivalent systems. OnStar has many features, including the ability to unlock vehicle doors and turn off the engine if, for example, the vehicle has been reported stolen. It also automatically detects if the vehicle is involved in a serious accident and, if so, it will contact the police. However, for this analysis, we are more interested in the monitoring and reporting capabilities of the system.

The regular OnStar service collects such data as tire pressures and odometer readings (from which it calculates oil life). It also monitors vehicles' engine and transmission systems, emission system, air bag systems, stability control systems, antilock braking systems, and the OnStar system itself. All these results are reported monthly by email to the owner. It also tracks whether warranties are due to expire. Turn-by-turn navigation and hands-free telephone services are optional.

Introduced for GM vehicles of the 2015 model year and later, the OnStar Smart Driver system collects additional information including time-of-day, speed during hard braking and hard acceleration, ignition on/off, time over 80 mph (miles per hour), distance travelled, fuel level, and idle time, from which the system calculates mph and mpg (miles per gallon). Driving activity reports are provided monthly. One goal of this system is to provide lower insurance premiums, although the opposite could happen if driving habits are not exemplary. The Smart Driver system also addresses privacy issues, such as whether illegal driving activities are reported to authorities, as with traffic-light camera systems.

3.6 Surroundings-to-infrastructure systems

In one sense, the difference between surroundings and infrastructure is just a matter of distance, with surroundings limited to, say, 200–300 meters from a vehicle, whereas infrastructure can cover areas of several square miles. However, the relationships between the two will depend upon how functions are divvied up among various systems. For example, notification of traffic congestion might come from the ecosystem, such as from Google, as described in [14], or from roadside or in-road sensors, which can either alert drivers through signs or alternatively transmit the information to infrastructure systems.

3.7 Infrastructure-to-ecosystem systems

While not usually considered by researchers, communications between local (municipal) traffic control systems will undoubtedly develop and, over time, may well exceed potential communications between the ecosystem systems and vehicles or surroundings. One reason for this is that infrastructure systems will both predict and control traffic, whereas vehicles, say, are not necessarily able to predict what the traffic control systems will do. This has the potential of being able to use real-time big-data predictive analysis and AI (artificial intelligence) methods to bridge across infrastructures and to improve local systems' ability to anticipate traffic patterns, particularly for journeys that cut across many local zones.

4 DATA COLLECTION, ANALYSIS, AND ACTIONS

The attributes and capabilities of systems can be itemized according to the functions of existing and potential systems, which are as follows:

- Data collection
- Analysis and reporting
- Recommendations for action
- Driver assistance
- Vehicle control

Table 2 lists various data items collected and shows whether the systems already exist or might eventually be realized. Many of these capabilities already exist for IV and V2E systems and some are expected in V2V, V2S, and V2I systems as well. It should be noted that the lists in the following tables are not exhaustive.

In Table 3, we indicate how the results based on the analysis of data collected are reported, such as via displays, sound, vibrations, etc.

In Table 4, we recommend how drivers should respond to the alerts that they receive. If the suggested actions are not followed, then alerts and recommendations will continue.

In Table 5, we show which driver-assistance capabilities have been, or are likely to be deployed. In these circumstances, the driver remains in full control of the vehicle, but is not only alerted to various steering and braking activities that should be done in response, but may also receive nudges to steer, brake, or, in the case of blind spots, not to change lanes.

In Table 6, we indicate areas in which IV systems might take over control of the vehicle, especially when the driver does not respond to alerts quickly enough to avoid accidents.

The above tables show that, for all five functions, many of the capabilities span several systems. It is here that decisions must be made as to where the capabilities should reside and, if they exist in more than one system, developers must determine which system is primary. For example, we show in Table 6 that automatic braking has already been deployed for IV systems, but that we can expect V2V and V2S systems to be able to communicate with vehicles' automatic-braking systems. The question arises as to which system decides if there are differences between the IV system, which gets its information from on-board sensors, and (say) V2V systems that base their decisions on the relative location and speed of two or more vehicles. From scanning the above tables, one can identify many such cases where system designers must resolve such conflicts unambiguously.

Table 2: Data-collection capabilities for existing and anticipated systems.

Function	Attribute/Capability/Metric	Existing (X) & Potential (P) Systems				
		IV	V2V	V2S	V2I	V2E
Collecting data	Time of day	X				X
	Speed	X	P		P	X
	Speed limits	P	P		P	P
	Speed over 80 mph	X				X
	Total idle time	X				X
	Mileage	X				X
	Steering	X		P	P	X
	Lane keeping	X		P	P	X
	GearForward, reverse	X				
	Hard braking, hard acceleration	X				X
	Ignition on/off	X				X
	Late night driving	X				X
	Battery voltage	X				
	Tire pressure	X				X
	Distance to objects in front	X				
	Engine fluid levels (gas, oil)	X				X
	Fuel use in MPG (miles/gallon)	X				
	Engine temperature	X				
	RPM (revolutions per minute)	X				
	Outside temperature	X		X		
	Rear view camera	X				
	Front camera (built-in, dash cam)	X				
	Total distance travelled	X				
	Trip distance travelled	X				
	Distance with remaining fuel	X				
	Vehicle location					X
	Central oversight and monitoring					X
	Possibility of ice	X			X	
	Upcoming traffic jams				X	X
	Possibility of upcoming collision	X	P			

5 PHASING OF SYSTEMS

Looking at historical and current efforts, we see that IV systems are generally well ahead of systems relating to surroundings, infrastructure, and the ecosystem. Due to such initiatives as the DARPA Grand Challenge [16], which began in 2004, the design and development of IV systems have been proven viable in many cases and vehicle manufacturers and software developers have already introduced subsystems, such as adaptive cruise control, into current models.

Table 3: Methods for reporting results of analyses.

Function	Attribute/Capability/Metric	Display (D), Sound (S), Radio (R), Vibration (V), Wireless (W), Email (E), Existing (X) and Potential (P) Systems				
		IV	V2V	V2S	V2I	V2E
Analysis and reporting of data	Navigation system					
	• Current location	D		P		W
	• Current direction	D		P		W
	• Distance to destination	D				W
	• Time to destination	D			D	W
	• Traffic congestion	D		P	D R	W
	• Time to location	D			D	W
	Time – Clock	D				
	Speed – Speedometer	D				W
	Need to obey speed limit	D P	P			
	Need to avoid collision	D S X				
	RPM – Tachometer	D				
	Total/trip mileage – Odometer	D				
	Moving outside of lane	D S V				
	Too close to vehicle in front	D S	P			
	Vehicle behind is too close	D S	P			
	Vehicle in blind spot	D S	P			
	Vehicle/person approaching when reversing	D S	P			
	Gears	D				
	Ignition	D				
	Battery:					
	• Voltage level (unless fully discharged)	D				E
	• Battery low (flashing side lights)	D				E
	Low tire pressure warning	D S				E
	Engine fluids and fuel:					
	• Current level	D				
	• Low fluid and fuel levels	D S				E
	• Fuel rate of use	D				
	• Estimated distance on remaining fuel	D				
	Temperature:					
	• Engine overheating	D S				
	• Outside temperature	D S				

(Continued)

Table 3: *(Continued)*

Function	Attribute/Capability/Metric	Display (D), Sound (S), Radio (R), Vibration (V), Wireless (W), Email (E), Existing (X) and Potential (P) Systems				
		IV	V2V	V2S	V2I	V2E
	In-vehicle warnings:					
	• ABS (automatic braking system) problem	D S				E
	• Engine service needed	D S				E
	• Other services needed	D S				E

Table 4: Recommended actions in response to alerts.

Function	Attribute/Capability/Metric	Display (D), Sound (S), Radio (R), Vibration (V), Wireless (W), Email (E), User Manual (M)				
		IV	V2V	V2S	V2I	V2E
Actions suggested to drivers	Navigation system:					
	Take preferred route, e.g. shortest distance	D S				W
	Make turn after a specific distance	D S				W
	Respond to revised instructions	D S				W
	Engine fluids and fuel:					
	Add fluids and fuel when levels are low	D S				E
	Replace fluids when recommended life met	D S				E
	Recharge or replace battery	D S				E
	Increase tire pressure	D S				E
	Reduce speed to limit	D S				
	Drive cautiously on ice	D S				
	Take vehicle for service	D S				E
	Engine overheated:	D S				
	Switch off engine	M				
	Add coolant	M				
	Check for coolant leaks	M				
	Return to lane	D S V	P			
	Reduce speed if too close to vehicle or obstacle in front	D S	P			
	Accelerate if vehicle to the rear is approaching too fast	D S	P			
	Do not switch lanes	D S V	P			
	Stop if vehicle/person approaching when reversing	D S V	P			
	Accelerate or brake if traffic lights about to change	D			P	

Table 5: Various driver-assistance methods in response to alerts.

Function	Attribute/Capability/Metric	Existing (X) & Potential (P) Systems				
		IV	V2V	V2S	V2I	V2E
Driver assistance	Aided steering to stay in lane	X	P			P
	Aided braking for obstacle ahead	X	P			
	Aided steering around obstacles	X	P	P		
	Aided response to vehicles in blind spots	X	P			
	Accelerating/decelerating – adaptive cruise control	X	P			
	Braking to obey speed limit	P				

Table 6: Existing and potential control systems.

Function	Attribute/Capability/Metric	Existing (X) & Potential (P) Systems				
		IV	V2V	V2S	V2I	V2E
Vehicle control	Steering	X				P
	Automatic braking	X	P	P		
	Accelerating/decelerating – adaptive cruise control, obeying speed-limit signs and the like	X P				P
	Turning off engine	X				X
	Keeping in lane	X				
	Unlocking doors remotely	X				X

On the other hand, infrastructure-related systems are mostly at the initial stages of development and are being tested in select areas.

Since there will be many interdependencies among all levels of systems, we must ask whether the IV system implementers are in a 'hurry-up-and-wait' situation. What appears to be happening is that vehicle manufacturers and software development companies are building overly complex IV systems to compensate for the lack of progress in external systems. While it may be acceptable to some players to rely totally on IV systems today, we are clearly heading towards a situation in which further progress will be stymied by limitations and deficiencies in V2V, V2S, V2I and V2E systems.

While there is general agreement that there are significant issues relating to the deployment of systems other than IV systems, many such systems face major challenges, as described in the following text.

6 CHALLENGES TO IMPLEMENTATION

There are many technical, structural, economic, and political challenges to implementing the various systems described earlier. High on the list is the determination of who is responsible for developing the systems and their interfaces and who is liable when something goes wrong. Appendix III of [7] shows the ratings by 21 experts of the challenges facing the deployment of V2V technologies. The biggest challenges were shown (in rough order of significance) to be:

- Establishment of a system management framework, with roles and responsibilities
- Technical challenges of developing V2V devices, driver–vehicle interfaces, and (especially) V2V safety applications
- Technical development of a data security system
- Accepted use of DRSC (dedicated short-range communications)
- Potential need for roadside equipment
- Public acceptance
- Human factors
- Liability issues relating to legal responsibility for crashes
- Deployment of devices and applications across enough vehicles and infrastructures to realize significant benefits
- Standardization to ensure interoperability among systems
- Acceptable end-user privacy
- Costs of deploying interconnected systems-of-systems

The results in [7] are heavily biased toward system security, which is a very valid concern and is being addressed by this researcher elsewhere. However, systems and their interfaces must first be designed for appropriate security controls to be established. Indeed, security requirements must be explicitly included in the general requirements, as recommended in [17]. Therefore, for the purposes of this article, systems challenges are mostly generalized to overall systems-of-systems.

It is interesting to note the predominance of concerns about establishing a structure that will govern the deployment of systems and assign roles and responsibilities, and various technical challenges. Standardization, privacy, and costs are seemingly less challenging. However, when it comes to surroundings and infrastructure, costs may be one of the more difficult issues to resolve since funding will likely come from government sources, which are not profit-driven. With respect to liabilities, the ruling on the above-mentioned case sided with Tesla Motors, which was not found to be liable [9]. This may indicate how such cases will be resolved in the future.

7 COMPLICATED AND COMPLEX SYSTEMS

Current and future IV systems, which are presented very well in [18] are, and will be, complicated, in that they may be difficult to comprehend taken together, but can generally be broken down into more understandable pieces. To quote from [19]: 'A car is not complex, just complicated'. However, as vehicles interconnect with one another, the resulting systems-of-systems will become complex since the behavior of combinations of these systems, which will be combined into vast systems-of-systems, will no longer be predictable, especially since they do not presently adhere to any predetermined standards across the automotive industry.

8 ADAPTIVE AND SELF-ORGANIZING SYSTEMS

Current intelligent transportation systems are essentially deterministic, even if they are interconnected and interoperate. They are more appropriately called 'expert systems'. They perform in predictable ways, which have been preprogrammed into the systems.

Looking forward, there is great interest in building systems that are truly adaptive, which is the basic requirement for AI systems. An excellent overview of the current status of self-driving cars and an interesting preview of what we might expect in the near future are given in [20]. With respect to training a vehicle to anticipate the unexpected, it is suggested in [20] that there are two ways to achieve this – either program in every possible eventuality, which is impossible with the current infrastructure, or teach a vehicle to learn and think for itself, which raises its own set of issues. Also in [20], Professor Philip Koopman of the National Robotics Engineering Center (NREC) at Carnegie Mellon University says that 'he worries the [automotive] industry is seriously underestimating how hard it will be to build innate safety features into artificially intelligent cars'.

In [21], there is a chapter on 'self-organizing traffic lights", which is particularly relevant to the situations discussed in this article. Gershenson [21] states that, as of 2007, mathematical and computational methods for controlling traffic lights did not consider the state of traffic in real time and were 'blind to "abnormal" situations, such as many vehicles arriving or leaving a certain place at the same time, such as a stadium after a match …' While optimization methods might give the best solution for a given configuration, it is asserted in [21] that an adaptive mechanism would perform better than optimization and that while each traffic light is 'unaware of the state of other intersections … [they] still manage to achieve global coordination'.

9 CONCLUSIONS

While the case for system integration among vehicles and their environment has been vehemently argued by many researchers and journalists, the mission to design and implement such complex systems-of-systems is fraught with technical and structural issues. There is clearly a need to bring diverse efforts together and encourage collaboration among players.

We have described the complex systems and interactions making up the current autonomous-vehicle situation and have indicated how that situation might develop over the short term. However, we are facing many challenges for developing and deploying long-term systems-of-systems and these must be addressed if we are to achieve the goal of completely safe and reliable autonomous ground vehicles and road systems. It is strongly recommended that a global facilitating group be created, made up of all interested parties, and that this group develop a generally accepted design that will result in safe and secure intelligent automotive systems-of-systems.

REFERENCES

[1] Kaplan, J., Roads that work for self-driving cars. *The Wall Street Journal*, updated 8 July 2016, available at https://www.wsj.com/articles/roads-that-work-for-self-driving-cars-1467991339 (accessed 9 February 2017)

[2] Valdez-Dapena, P., Volvo promises deathproof cars by 2020. *CNN Money*, 21 January 2016, available at http://money.cnn.com/2016/01/20/luxury/volvo-no-death-crash-cars-2020/ (accessed 9 February 2017)

[3] Petroski, H., Why cities aren't ready for the driverless car. *The Wall Street Journal*, updated 22 April 2016, available at https://www.wsj.com/articles/why-cities-arent-ready-for-the-driverless-car-1461550001 (accessed 9 February 2017)

[4] Abuelsamid, S., Audis will talk to some traffic signals, kicking off vehicle to infrastructure communications. *Forbes*, 16 August 2016, available at http://www.forbes.com/sites/samabuelsamid/2016/08/16/2017-audis-will-talk-to-some-traffic-signals-kicking-off-vehicle-to-infrastructure-communications/#6fbd0c663efd (accessed 9 February 2017)

[5] United States General Accountability Office, *Vehicle Cybersecurity: DOT and Industry Have Efforts Under Way, but DOT Needs to Define Its Role in Responding to a Real-world Attack. GAO-16-350*, March 2016, available at http://www.gao.gov/assets/660/658709.pdf(accessed 9 February 2017)

[6] Walker, J., *ITS America 2016: FWA Vehicle-to-Infrastructure (V2I) Deployment Guidance and Products*, Federal Highway Administration, Washington, DC, 2016, available at http://www.its.dot.gov/pilots/pdf/ITSA2016_v2iGuidance_Walker.pdf (accessed 9 February 2017)

[7] United States General Accountability Office, *Intelligent Transportation Systems: Vehicle-to-Vehicle Technologies Expected to Offer Safety Benefits, but a Variety of Deployment Challenges Exist. GAO-14-13*, November 2013, available at http://www.gao.gov/assets/660/658709.pdf (accessed 9 February 2017)

[8] Neil, D., Mercedes E400 review: Making better drivers of us all. *The Wall Street Journal*, 19 January 2017, available at https://www.wsj.com/articles/2017-mercedes-benz-e400-review-making-better-drivers-of-us-all-1484850721 (accessed 9 February 2017)

[9] Boudette, N.E., Tesla's self-driving system cleared in deadly crash. *The New York Times*, 19 Jan. 2017, available at https://www.nytimes.com/2017/01/19/business/tesla-model-s-autopilot-fatal-crash.html?_r=0 (accessed 9 February 2017)

[10] Ross, P.E., Why can't government run vehicle-to-vehicle communications? *IEEE Spectrum*, 20 August 2014, available at http://spectrum.ieee.org/cars-that-think/transportation/infrastructure/why-cant-the-government-run-vehicletovehicle-communications (accessed 9 February 2017)

[11] Ross, P.E., Europe's smart highway will shepherd cars from Rotterdam to Vienna. *IEEE Spectrum*, 30 December 2014, available at http://spectrum.ieee.org/transportation/advanced-cars/europes-smart-highway-will-shepherd-cars-from-rotterdam-to-vienna (accessed 9 February 2017)

[12] Page, P., States wire up roads as cars get smarter. *The Wall Street Journal*, 2 January 2017, availabe at https://www.wsj.com/articles/states-wire-up-roads-as-cars-get-smarter-1483390782 (accessed 9 February 2017)

[13] Geller, T., Car talk: vehicle-to vehicle communication is coming, are we ready for it? *Communications of the ACM*, **58**(3), pp. 16–18, 2015. https://doi.org/10.1145/2717177

[14] Volvo – Car-to-car communication. *YouTube*, https://www.youtube.com/watch?v=O0xBCSmOfmE (accessed 9 February 2017)

[15] Brindle, B., How does Google Maps predict traffic? *How Stuff Works*, 31 October 2014, available at http://electronics.howstuffworks.com/how-does-google-maps-predict-traffic.htm (accessed 9 February 2017)

[16] Defense Advanced Research Projects Agency (DARPA), DARPA Grand Challenge: Frequently Asked Questions, DARPA Web Site, Arlington, VA, 10 July 2006, available at http://archive.darpa.mil/grandchallenge/docs/Urban_Challenge_General_FAQ.pdf (accessed 9 February 2017)

[17] Axelrod C.W., *Engineering Safe and Secure Software Systems*, Artech House: Norwood, MA, 2012,

[18] U.S. Department of Transporation – Vehicle to-vehicle communication. *YouTube*, available at https://www.youtube.com/watch?v=POcQUTlOvZs#t=591.2487048 (accessed 9 February 2017)

[19] Kaisler, S. & Madey, G., Complex adaptive systems. *HICSS-42*, Hawaii, 5 January 2009, available at https://www3.nd.edu/~gmadey/Activities/CAS-Briefing.pdf (accessed 9 February 2017)

[20] Plungis, J., Driving into the future. *Consumer Reports*, **82**(4), pp. 10–16, 2017.

[21] Gershenson, C., *Design and Control of Self-organizing Systems*, Ph.D. Dissertation, Vrije Universiteit Brussel, 2007, available at http://cogprints.org/5442/1/thesis.pdf (accessed 9 February 2017)

MULTI-AGENT TASKS SCHEDULING FOR COORDINATED ACTIONS OF UNMANNED AERIAL VEHICLES ACTING IN GROUP

PETR SKOBELEV[1], DENIS BUDAEV[2], ALEKSEY BRANKOVSKY[2] & GEORGY VOSCHUK[2]
[1]Samara National Research University, Samara State Technical University, Samara, Russia.
[2]Smart Solutions, Ltd, Samara, Russia.

ABSTRACT

This paper discusses hardware and software prototype of multi-agent system for unmanned aerial vehicles (UAVs) group action planning. We describe the approach for system implementation as a whole and software agents within the system. The aim of current and future developments is creation of complex scientific and technical solutions for coordinated planning and actions management of heterogeneous UAV groups in real time.

Keywords: adaptability, coordinated control, drones, dynamic rescheduling, intelligence, multi-agent systems, real time, UAV, Unmanned aerial vehicle

1 INTRODUCTION

The use of classical optimization methods is not always effective for complex multifactorial problems, especially with high dynamics of events affecting the results of the planning. Reconciliation in a distributed computing environment also requires special approaches. An example of such complex problems is coordinated management of a group of unmanned aerial vehicles (UAVs) in real time.

Unmanned aerial vehicles (UAVs or drones) are widely used in various sectors of economy. Main advantages of using drones are achieved when used in critical missions that require response in the shortest possible time. For example, UAVs can search for survivors at emergency scenes, which provides quick access to areas where human access is impossible or dangerous. However, to ensure the implementation of the assigned tasks in the shortest possible time and improve the quality authors consider using several drones acting together as a coordinated group. Such a group of drones with a joint goal should coordinate and reconcile mutual tasks. As a result, a group of UAVs acting together will be able to perform tasks faster.

However, there is an issue with planning tools for real-time control of UAVs groups, especially in unforeseen circumstances. For example, there is lack of software, which can manage UAVs group in case of failure of a UAV, addition of a new area, or addition of a new UAV in a group. Most of the existing software solutions are designed for planning one autonomous UAV. On the other hand, drone group management software would use available UAVs resources in the most efficient way (battery or fuel supplies, time resources, hardware/computing resources of UAVs). In addition, quick distribution of tasks between the drones during mission execution would reduce total time.

For effective management of resource allocation (incl. drones and their subsystems) it is expedient to use scheduling systems. Nowadays these kinds of systems use the following methods of complex problem solving:

- traditional methods of optimization and linear programming in the area of mixed real-valued, integral-valued and logical variables, the improvement of precise methods of tasks solving, such as 'branch and bound' methods, nonlinear programming methods, methods of constraint programming [1];

DOI: 10.2495/DNE-V13-N1-39-45

- greedy algorithms, based on heuristic business-rules for specific subject areas;
- artificial intelligence methods, the use of neural networks and fuzzy logic;
- metaheuristics (local search, Tabu Search, GRASP algorithms) [2];
- bio-inspired methods: Ant Colony Optimization (ACO), Artificial Bee Colony (ABC), Bio Inspired and a similar methods, as well Simulated Annealing (SA), Monte-Carlo method and some others [3].

Many scheduling systems are based on centralized and deterministic principles. However, distributed coordination in networks of dynamic agents has attracted an interest of numerous researchers in recent years. As shown in Refs. [4, 5], the multi-agent technology methods are the most promising and appropriate for the resources allocation algorithm design. The following paper describes the practical experience of using multi-agent technology methods for developing a system for UAVs group management. We describe task formulation for terrain survey with UAVs group and decentralized multi-agent, multi-criteria task scheduling solution process. We also describe the results of the software and hardware implementation of the system along with some experimental and real flights results.

2 PROBLEM STATEMENT

Let us consider problem formulation with the following initial data:

- area for survey mission that is limited in size and may be identified by boundary points coordinates;
- topography information about the area (terrain heights);
- collection of available UAVs, which may be included in a group;
- technical specification of UAVs and equipment - ranges of available speeds and altitudes, battery capacity, charging time, camera information.

One of the main features of proposed UAVs group management approach is the ability of UAVs to communicate over the wireless network. In addition, each UAV can fly in standalone mode along the path trajectory identified by checkpoints, and each unit can 'coordinate' actions with other devices by sending and receiving data messages via wireless communication channels.

3 UAV GROUP MANAGEMENT

UAVs have expanded the scope of their application in recent years quite extensively. For example, rescue services use UAVs for forests monitoring, fires search. UAVs re successfully used in the search and rescue missions. UAVs can help solve problems in the areas with difficult or dangerous access for humans.

Using UAVs increases the effectiveness of rescue operations because they are mobile and require less time to prepare for the flight and launch (when compared to the large aircrafts). UAVs are indispensable for creation of accurate digital models of territories with a dense arrangement of objects and buildings.

Most of the existing software solutions are designed for planning and management activities of alone acting UAV. However, some missions require rapid tasks distribution among available drones. It is critical to calculate individual routes and actions of each UAV and reconcile resulting routes especially in high dynamics of events affecting the results.

Ordinary methods are not always effective for solving complex multifactorial problems, especially in high dynamics in distributed environment. One approach to solving the problem of resource management involves the use of multi-agent technology (MAT).

As part of the multi-agent approach, each active entity type within a solved problem is represented by a software 'agent', formalizing the logic and needs of the entity. All UAVs have on board a separate single-board computer with MAT software, so UAVs can share information wirelessly. Scheduling in this case is a process of negotiation between these agents in order to determine the compromise resulting plan.

For UAV group management we introduce the concept of observation square. The surveying area is divided into a finite amount of observation squares according to characteristics of the UAVs and their equipment. For example, for the UAV with camera sensor width 4.55 mm, the focal length 3.61 mm, the flight height 119 m (FAA restricts flights for small UAVs above 400 feet altitude), the image width and height 4000 and 3000 pixel, respectively, we expect ground sampling distance (GSD) of 3.75 cm/pixel and image footprint on the ground 150 m. Therefore, for the area of 32 x 32 observation squares we expect the total area of 23.04 square kilometers. Thus, there is a survey mission with 1024 sub-tasks for available amount of UAVs.

Each observation square is associated with its sub-task to perform. Each sub-task has a timestamp of UAV flight and its format is hh:mm:ss. The timestamp is renewed at the next UAV flight over the observation square. Considering these facts, UAVs agents can track how long the square was without supervision. It is important for the patrolling mission planning.

After operator sets mission parameters (sets the boundary points of observation area, selects a mission type, for example, 'single flight over area' and determines the UAVs in-group), all sub-tasks become available for scheduling and are transmitted to UAVs matching mechanism. This mechanism is responsible for the initial distribution of individual sub-tasks (observation squares) between individual UAVs of group. In fact, the mechanism realizes clustering to aggregate observations squares.

After the initial clustering each UAV agent receives its individual area within the corresponding cluster, that is, some set of observations squares, in which the agent has to build a flight path to perform UAV area supervision.

The process of flight path calculation is performed by UAV agent. UAV agent analyzes the list of available sub-tasks and evaluates alternatives sets of paths based on the set of planning criteria. Final list of planning criteria is primarily determined by the type of the mission.

Authors consider that it is important to build such multi-agent scheduling algorithms that UAVs have the least possible number of turns and backward turns. To solve this issue by analogy with the method proposed in [6] for generating a path for a single UAV, we propose an algorithm modification for UAVs group.

During the planning process and performance of in-flight sub-tasks each UAV can reallocate sub-tasks according to agent key performance indicators. This approach allows controlling sub-tasks distribution and balancing in real time, and unlike similar methods of path forming [6–9], it is originally designed to increase total system performance, resources utilization and reduce execution time. Agent key performance indicators (KPI) are applied to evaluate and compare different options of sub-tasks distribution based on criteria of UAV agent satisfaction. For prototype system two criteria have been chosen to calculate the agent satisfaction:

1. Criterion of area observability. This criterion depends on sub-tasks squares observability. The meaning of this criterion is to enforce agents selecting those observation squares which have been most unexplored in the previous time intervals. UAV agent satisfaction depends on the time during which the square remained without supervision, that is, from the time during which no UAVs of a group were flying over a square for observing.
2. Criterion of total UAVs path distance. The meaning of the criterion is to enforce agents selecting a set of squares which lies on the same line. This provides efficient use of UAVs resources - minimum number of turns on the path and minimum distance without observing a square. The criterion defines satisfaction of UAVs agents from alternatives of observation squares selecting. UAVs agent KPI are higher when his path contains a set of observations squares lying on a straight line. The attractiveness of selected squares set for UAVs agent is higher, the higher the ratio of paths length on observed squares to the total path length.

The overall system satisfaction at a certain stage of work is defined as the sum of all UAVs agents satisfaction on the criterion of area observability and on the criterion of total UAVs path distance considering criteria weights (sum of the weights is always equal to 1).

$$KPI_{system} = k_{area\ observ.} * \sum_{i=1}^{N} KPI_{i\ area\ observ.} + k_{path\ dist.} \sum_{i=1}^{N} KPI_{i\ path\ dist.} \tag{1}$$

During coordination of the flight plan in the group, each UAV provides a set of possible options for changing position. Each of these options is characterized by satisfaction indicators for UAV (observability and path distance). The overall system satisfaction at a particular planning stage is defined as the sum of KPIs of all agents, considering criteria weights. The higher the total KPI of the system, the better the quality of solution in the context of the selected criteria.

Thus, the total flight plan is a set of options for all devices of a group, one approved option from the list of alternative options for each UAV, which maximizes KPI of the system.

Realized dynamic scheduling mechanism provides adaptive relocating of individual observation squares between UAVs in the group to minimize difference between completion times between different UAV devices.

A process of sub-squares relocating between two UAVs is performed as follows:

1. During flights UAVs exchange data about execution time forecasts.
2. UAV detects a significant difference (parameter) between its completion time forecast and the time of another UAV.
3. UAV agent calculates how many sub-tasks to transfer from or to other UAV to reduce the difference between completion times.
4. UAV agent sends a request to reallocate the calculated number of sub-tasks.
5. Sub-tasks are reallocated or the agent receives a rejection (for example, if UAV has already transferred some of the squares to another third UAV).

Thus, adaptive balancing mechanism ensures that the mission is performed in the shortest time even in cases of UAVs various performance and unplanned events.

4 RESULTS

Authors have completed a series of experiments to research the system performance. The input data contained survey area, collection of available UAVs (software-hardware simulations with PX4 flight controller boards). The total amount of observation squares was 1024. The main purpose of experiments was to evaluate effectiveness of sub-tasks distribution by UAVs of the group, assess formed UAV flight plan, estimate time required for planning and preparing flight plans, and assess systems reaction to tasks changing during planning. During tests, we launched scheduling process for different amount of UAVs in group and measured parameters of system performance. Results of experiments are shown in Table 1.

- Firstly, total time spent on tasks performing directly depends on the number of observation squares in the flight mission and the number of UAVs devices involved in the planning process. With the same number of squares the more devices are involved in the planning process, the faster the system can issue a total flight plan. This is explained by the distributed nature of planning mechanism, which allows use of all resources of computing devices in the planning process, which reduces the time for preparing the final total flight plan. Scheduling for a single-working UAV may take longer than for united group of devices, since all subtasks have to be consistently rated by one device.
- Secondly, the greater the number of UAVs involved in mission, the less time is required to perform all tasks of the total flight plan. This is easily explained by the fact that each of UAVs works in its responsibility area concomitantly coordinating work plans and levelling load with all UAVs from a group using the load balancing mechanism that reduces the total time of full mission execution.
- Thirdly, the system reaction time to changes in environment is less when a smaller number of UAVs is affected by this change, because coordinating all changes requires negotiation between all agents affected by these changes.

Also overall efficiency of the system is characterized by the indicator of total KPI of the system which in turn is determined by satisfaction level of each UAV agent. Generally, the KPI value for a single-working UAV will be higher than the overall KPI value for a group of coordinated devices which are forced to compete and negotiate, by compromising. In other words, KPI index is only a signal about necessity of plan optimization during system performance and these indicators only can be compared with similar indicators of similar UAVs groups, for example, similar UAVs types and number of devices.

Table 1: Summary results of experiments.

Indicator	Group of 3 UAVs	Group of 4 UAVs	Group of 10 UAVs	Single UAV
Time for full mission planning, seconds	62	50	21	165
Forecasted time for mission performance, minutes	404.23	304.23	145.33	1205.13
Reaction time to tasks addition (time of sub-tasks relocation between UAVs), seconds	9	15	13	6
Overall system KPI (satisfaction of all agents) after planning, %	78	79	68	95

Multi-agent planning methods provide the ability to manage a group of UAVs and monitor and adjust performance of the group through agent key performance indicators and criteria.

5 CONCLUSION

Software systems for management and control of robotic devices including united UAVs groups are actively developed at the present time [10, 11]. A promising direction for developing such systems is extending their functionality by developing technical solutions using methods and tools of coordinated planning of actions in groups of robotic devices performing common tasks. Furthermore, for successful using of such systems, their functionality should allow for adjustment of formed execution plan in changing environment, including respond ability to un-foreseen situations through relocating subtasks between individual devices of group. In case of changes in the environment, their functionality must allow for adjustments of the executable plan.

According to the authors, distributed planning via interactions between UAVs has important advantages over centralized planning. Here is just a small number of these advantages:

- faster reaction to external changes than in centralized planning systems;
- simple change of execution plan (inclusion/exclusion of UAVs, replacement of faulty UAV);
- use of various models and types of UAVs.

Distributed multi-agent planning is organized through communication between individual modules on each UAV, connected to the peer-to-peer network. These modules are implemented on the base of single-board computers (such as Raspberry PI, Arduino). The developed system allows for planning and correcting actions of the UAV group when some events are triggered through interaction of individual computing modules. Therefore, tasks are solved regardless of the number of used UAVs and taking into account the amount of criteria introduced into the model which influence the overall performance.

The tasks executed using drones, which are most demanded by the market, include:

- fire and rescue operations;
- area monitoring and inspection for protection and control;
- evaluation of vegetation indexes (such as NDVI) for agriculture;
- photo and video shooting of moving objects in the spheres of leisure, tourism, entertainment.

When you create a new hardware systems designed to solve problems of group control, the most expensive stage is software development [12]. Versatility of proposed methods and their independence from the context of the problem will reduce the time for deployment of new hardware systems and reduce costs of their creating by minimizing development time of software solutions. Thus, software and hardware complexes of the new generation are composed of various models and types of UAVs capable to interact and work together and may be designed and implemented in the shortest time. Multi-agent scheduling technology is applicable to a wide variety of practical tasks in various areas of economy and industry including UAVs group management in real time.

ACKNOWLEDGMENTS

This work was supported by Russian Foundation for Basic Research, project number 16-01-00759.

REFERENCES

[1] Michael, L., *Pinedo scheduling: theory, algorithms, and system*, Springer, New York, p. 673, 2008.

[2] Vos, S., *Meta-heuristics: the state of the art in local search for planning and scheduling*, A. Nareyek (Ed.). Springer-Verlag, Berlin, pp. 1–23, 2001.

[3] Binitha, S. & Sathya, S.S., A survey of bio inspired optimization algorithms. *International Journal of Soft Computing and Engineering* (IJSCE), **2**(2), pp. 137–151, ISSN: 2231-2307, 2012.

[4] Rzevski, G. & Skobelev, P., *Managing complexity*. WIT Press, Boston, 2014.

[5] Skobelev, P., Multi-agent systems for real time adaptive resource management. In *Industrial agents: emerging applications of software agents in industry*. Paulo Leitão, Stamatis Karnouskos (Ed.). Elsevier, Amsterdam, pp. 207–230, 2015.

[6] Santamaria, E., Segor, F., Tchouchenkov, I. & Schoenbein, R., Rapid aerial mapping with multiple heterogeneous unmanned vehicles. *International Journal on Advances in Systems and Measurements*, **6**(3–4), pp. 384–393, 2013.

[7] Di Franco, C. & Buttazzo. G., *Energy-aware coverage path planning of UAVs*. Autonomous robot systems and competitions (ICARSC), 2015 IEEE International Conference, pp. 111–117, 2015.

[8] Kamrani, F., *Using on-line simulation in UAV path planning*. Licentiate Thesis in Electronics and Computer Systems, KTH, Stockholm, Sweden, 2007.

[9] Ergezer, H. & Leblebicioğlu, K., 3D path planning for multiple UAVs for maximum information collection. *Journal of Intelligent & Robotic Systems*, **73**(1–4), pp. 737–762, 2014.

[10] Baxter, J.W., Horn, G.S. & Leivers, D.P., Fly-by-agent: controlling a pool of uavs via a multi-agent system. *The 27th SGAI International Conference on Artificial Intelligence*, **21**(3), pp. 232–237, 2008.

[11] Koo, T.J. & Shahruz, S.M., Formation of a group of unmanned aerial vehicles (UAVs). *American Control Conference* (ACC), **2001**, pp. 69–74, 2001.

[12] Austin, R., *Unmanned aircraft systems UAVs design, development and deployment*, 1st ed. Wiley Aerospace Series, United Kingdom, pp. 221–226, 2010.

DESIGN PROCESS AS COMPLEX SYSTEM

R. BARELKOWSKI
West-Pomeranian University of Technology, Szczecin, Poland.

ABSTRACT

Can a design process be complex system? Can it fulfill various criteria related to complexity, while its goals are, usually, temporarily defined and the process itself is expected to provide particular solutions transferable into physical volumes and solid environmental components?

It is apparent that the majority design cases do not follow traits and requirements of complexity, but this limitation seems to be related to natural tendency of simplification within architectural routines. Particularly public works, significant for the community, require the approach broadening the scope of understanding of spatial phenomenon, its role and its composition as a result of various programmatic, ideological, formal, and engineering aspects, embedded in complexity theoretical background.

Seven principal components of complexity, given by Rzevski and Skobelev, are more or less explicitly or implicitly present in design practice, and in particular, in design process: connectivity, autonomy, emergence, non-equilibrium, nonlinearity and self-organization. The aspect of evolution is the least apparent and there are significant limitations to what can be achieved there, mostly the process can evolve, while designed substance rarely can follow in the same flexible manner. In the paper I will argue that approach related to complexity is the general mode of architectural design, simplified in many cases due to human inclination to reduce the number of simultaneously processed problems and usually resulting in some design flaws or failures. This complex structure of design process, exemplified in the paper as a particular research case – the process for local cultural center – is the basis, which can be furthermore simplified, contrary to the idea that it is more sophisticated, non-standard approach. Working not only with the client, but with various types of users is a typical architectural condition, implementing significant constraints and at the same time forcing multiple organization arrangements within the process. The case will provide the platform to discuss broader idea of design as complex environment for the architect.

Keywords: architectural design process, complex system, complexity, Meta-Design

1 INTRODUCTION

The architectural design process is a multifaceted phenomenon reflecting plentitude of diverse disciplines and aspects of human life projected onto human organization of their environment. Interpretations of this phenomenon are countless and discussing of this issue inevitably encourage the narrowing of the scope of any research conducted in order to understand its principles, even if. It is necessary to define the aim of this particular paper being a prolegomena for understanding the design process as a decision-making process immersed in the framework disclosing the complex nature of design. Therefore I will advocate to understand the architectural design process as complex system, which most often is reduced to its simplified form due to practical constraints of massive and multiple use, frequent application devoid of an effort of understanding reasons and motivations other than those required to straightforwardly fulfill the task at hand. I argue that regardless of how profound the impact of complexity theory has on architecture in context of critical analyses elements of architectural epistemology have inherent components of complex system seen in every design process.

The scientific approach to architectural design dates back to general engineering design research, and therefore it was built on foundations laid by Asimov *et al.* [1], Krick *et al.* [2], to name the few. Starting from simple, linear concepts of the advancement of the process like Asimov's model [3] or interdisciplinary depictions like in case of Krick [4], researchers

© 2018 WIT Press, www.witpress.com
DOI: 10.2495/DNE-V13-N1-46-59

attempted to delineate the course of design which, however, was incompatible with the discipline of architecture mainly because of the fundamental difference described by John Zeisel. Zeisel noticed that architectural goals, reflecting disciplinary problems, are established in completely different manner, than typical engineering goals, and to put it briefly that instead of seeking optimization architects seek for imprecise state of acceptance [5]. The observation came as a conclusion and criticism of attempts to use algorithms as basic, fixed building bricks of architectural practice, both to determine the solutions, and to determine the organization of design process. It was also connected strongly to the realization how important social issues, not to mention cultural issues, are in every architectural work.

2 ARCHITECTURAL COMPLEXITY – A BRIEF PANORAMA

Historically intuitive design dominated and still is the most often form of architectural practice. However intuitive mechanisms are insufficient when it comes to managerial abilities combining, for example, technical, technological, aesthetical, social and economical aspects. Every design process in architecture is a reflection of social life and at the same time an attempt to process various types of data as well as transform physical components in order to improve human-centered environment, and relying on gradual perfectioning of design skills, on context-driven ad hoc adjustments of already established routines is usually far from efficient. While architectural constructs – particular themes – may be less or more complex, and therefore both guidelines and recipe to tackle the problem should refer to general methodologies, encompassing all possible or at least majority of cases. Thus, research efforts are focused on finding universal understanding of architectural design process and it requires often overlooked concept of holistic approach to design, which so aptly Christopher Alexander refers to, when depicting his original contribution to design theory from the 1960s [6]. The most significant goal of architecture is to respond to or to extend rich and multilayered reality, in which physical volume of architecture is merely a vessel for both human ideas and human pursuits connected with functional aspect. Since 1980s design process models evolved and spread depicting different valid intuitive or even purely technical (organizational) concepts like Robert Gutman's [7] or recently Lorraine Farrelly's [8], but at their best they are fragmentary and therefore significantly limited in their application.

Architectural science postulated more thorough and more systemic approach to the issue of the process. The introduction of systemic approach reaches as early as 1957 when Allen Newell, John Shaw and Herbert Simon presented their work on information processing theory, which was later respectively applied to architecture [9]. Within the field of architecture, however, system appeared too rigid, and some of principles, like eight implications of the concept of a system collected by Arnold Benjamin Handler, didn't correspond to specific architectural problems [10], but at the same time some of these suggest the potential, later developed as part of so-called environmental model of design advocated by Jon Lang [11] and later on complexity approach. As Antoine Picon writes in his essay on continuity, complexity and emergence, the term complexity may be used in multiple ways, and his own thoughts concentrate on the digital design as an evolved form of contemporary, most recent background for ideas in architecture [12]. On the one hand, the philosophical discourse in architecture approved digital technologies as a source of new and adequate language, and therefore allowing to merge abilities offered by the new media with flourishing creativity opposing traditional ideals of pursuing perfection in architecture – instead responding to virtualization and to blurring of borders between the real and the simulation. On the other hand, Picon confirms that complexity can go further, implicitly referring to increased diversity and multi-threadedness of the environment with its split, yet mutually supplementing counterparts of the real and the virtual, posing more challenges for architects [13]. Little, however, we may find there regarding the role of the architecture – and this is where Alexander's understanding of architectural complexity proves its validity [14]. What is really important about Alexander's idea of complexity implementation is in the approach to holistic view of architecture, in integration between objective, objectivized, and subjective components of the design process. Contrary to self-contained problems and resultant processes, architects are expected to respond to real world even if important parts of this reality is virtual. Complex system is too valuable to employ it for fancy, theoretical deliberations, instead of offering visual manifestation instead of, what Michael Mehaffy and Nikos Salingaros define as connective and coherent complexity [15]. Both researchers see complexity as a platform to successfully reflect and address essential problems of architecture – for people.

3 COMPLEXITY MODEL IN ARCHITECTURE

Diverse concepts of architecture, or particularly architectural design process, link it to complexity theory primarily through two secondary level ideas – first, the understanding of architecture as immaterial environment constructed by people for themselves, containing all, even unintended elements, and second, the process and all its constituents and stimuli. One can see here a clear pattern: the reality and its preceding moderation. Within complexity architecture exists as rich, complex phenomenon, in which physical volumes are merely vessels for socio-cultural contents. It respects architecture as what defines physical components to organize, but these volumes and delimitations are nothing but results, symptoms of social and cultural processes. The architectural design process is aimed at transforming observed reality giving it an altered state, an improvement. Therefore elements of analysis, observation, comprehension come along with carefully constructed prediction. The process, even implicitly, aspires to provide positive change – more as a program suggesting appropriate steps to be undertaken in order to receive planned status.

The presence of prediction assumes uncertainty – in result implying necessity to contain unplanned, emergent phenomena or issues which may and will surface during either planning, designing or executing, occupying. What's more, design cannot be hierarchical – and it's not the issue of in-processual loops, but overall structure – and it also cannot be ahierarchical. It is clear that architect is responsible for building the system of criteria for every project, but this system must be flexible and sometimes must be rebuilt from scratch, even several times, within particular task. We may conclude, therefore, that architectural projects, or more precisely design processes, tend to become compatible with complexity desiderata.

Seven principal components of complexity were gathered by George Rzevski and Petr Skobelev: connectivity, autonomy, emergence, non-equilibrium, nonlinearity, self-organization and co-evolution [16]. It is hard to discuss the transformation of reality that constitutes architectural environment and how architectural profession is performed, but it is worth mentioning that many of complexity criteria respond to increased consciousness of customers, users of space, developers and clients from public sector – the architectural task, which was once focused on providing shelter or exposing social role, nowadays fulfills all these roles extended with many more concepts like the issue of comfort (reflection of lifestyle), safety, architectural semantics, programming spaces for the use of groups or congregations [c.f. Table 1]. The example of such was delivered by Rena Upitis in her study of school designs [17], and complexity evidently plays significant role in large scale architectural designs,

Table 1: Criteria of complexity and their application in the field of architecture.

Complexity criteria	Architecture	Architectural design process
Connectivity responds to emergence connects directly to autonomy contributes to nonlinearity, self-organization and co-evolution	Various components of actual spatial program – and in result spaces themselves, whether in urban or architectural scale, collectively and dynamically negotiate its role	**Intradisciplinary and interdisciplinary networking** general issues influence detailed considerations, detailed decisions affect principles, integration, interdisciplinary or transdisciplinary perspective is applied to the process
Autonomy responds to emergence connects directly to connectivity contributes to nonlinearity and self-organization	Various components of actual spatial program – and in result spaces themselves operate independently	**Ability to accommodate exclusive** specific branches and themes are independent and may conditionally operate regardless of other branches / themes as long as other criteria do not trigger reconfiguration
Emergence (rarely) responds to self-organization connects directly to non-equilibrium contributes to connectivity, autonomy, nonlinearity and co-evolution	Socio-cultural events or interactions emerge in configured space (environment), unpredictable use appears	**Inclusion (responsive) mechanisms** acknowledgment of unpredictable decision-shifting (sudden) factors / agents; recalibration of the remaining process (or extension of the process), generation of self-induced alterations to process
Non-equilibrium responds to emergence and co-evolution connects directly to self-organization contributes to connectivity and nonlinearity	Any arrangement between architectural components is unstable; even if it tends to balance, this state can never be effectively acquired	**Permanent fluctuation (flexible structure)** the process tends to reflect architect's personal evaluation system; however, it fluctuates constantly in response to new criteria or new relationship between criteria
Nonlinearity responds to emergence, non-equilibrium and self-organization connects directly to connectivity and autonomy	Continuity is represented in space by spatial-temporal sequence, but this continuity is defined by random events and drivers, potentially re-orienting the use of architectural/urban space	**Open architecture of the process** design process is performed in parallel network threads or cells and there are no sequential steps (but for contractual purposes)

(*Continued*)

Table 1: (*Continued*)

Complexity criteria	Architecture	Architectural design process
Self-organization responds to emergence and non-equilibrium connects directly to connectivity and autonomy contributes to nonlinearity and co-evolution	Architectural/urban space organizes its performance independently from its creator or even community that uses it; reconfiguration, repurposing is perpetual	**Permanent reconfiguration (flexible structure)** system of criteria for design as well as any relationship between factors / agents may be altered and constantly remain open to further re-organization depending on validity or applicability of current one; the structure is permanently adjusting itself
Co-evolution responds to emergence contributes to connectivity, autonomy, nonlinearity and co-evolution	Various elements of space evolve in different rate and diverse directions, yet coordinate its interconnected-ness	**Permanent evolution of contents of the process (flexible structure)** responsive modules within the process

extending the notion of complexity used by Esra Bektaş, particularly in case of large public projects [18] in which uncertainty and resultant risk of failure are increased.

Rzevski and Skobelev bring forth additional issue – significant question of adaptability being a primary requirement of the system. This is clearly the case in almost all architectural tasks, and therefore, it qualifies to define the generic framework of design process. Making design process an adaptive system is requested because of multitude of factors – customers' alteration of expectations or criteria, responses to availability of particular solutions, dealing with substitutive material, technological or spatial solutions, changing budget constraints, among others. Rzevski and Skobelev mention real-time decision making, delayed commitment, minimizing consequences of disruptions, distributed decision-making, anticipation, experimentation and learning from experience [19].

Few of these adaptability traits urge for comment. Anticipation, or in more elaborated cases analysis-based prediction is not so obviously embedded into architectural design process. However clearly this element should be acknowledged in every design task – let us consider one of the simplest examples possible, which is detached family house. The anticipation is related not only to what customer wants today and tomorrow (e.g. children getting mature and leaving), but also what may happen to customer or their relatives (e.g. accidents, unpredictable disabilities). Naturally, in many components of design process this anticipatory aspect reflects varied level of requested ability to determine future performance of architecture, and the architect should be constantly aware to react and adjust the course of design respectfully. Learning from experience may be extended here by the inclusion of learning organization levels. As suggested by Robert Barelkowski, third level organization of learning [20], which is learning from learning supplements learning from designing, which is implied 'learning from experience'. Additional learning structure introduces axiological perspective, responsive methodological approach, adjustable value systems, and multi-disciplinary imports interwoven into rich tapestry of expanding professional knowledge management. Experimentation is the next topic which is altered by the application within the framework of complexity. It implies fragmented and dynamic experimenting with architectural concepts. Instead of organizing and subordinating the entire process to allow for design as experiment, the experimentation is divided into autonomous cells and launched contextually regarding the perspective of its necessity and efficiency. It is reflected in 'designing' the framework, being able to accommodate separate experimental activities performed simultaneously by different members of design team, in some circumstances even in purposefully uncoordinated manner. Distributed decision-making and real-time decision making are closely connected and appear in design process very overtly. Clear example of this feature may be given in the way installation or engineering designs may affect architectural part of solution. Certain constraints are driven by availability of technologies and services and may vary between what is available in time of particular decision is (initially) made and changes that are imposed by shortage of suggested conclusion. The progress of the project requires instant reaction and similar behavior is expected when conflict (e.g. aesthetical, between architecture and HVAC systems) appears. Delayed commitment is somewhat problematic in architectural implementation. On the one hand actually there is a suspension of final decision, but on the other hand design must deal with solid pieces of information, and therefore often working assumptions are taken into account. In the process this 'suspension' is counterproductive, and therefore, often replaced by multi-variant open solution, following temporarily particular concept with preparation for absorption of altered one or even rejection. In the latter case acceptance for rejection perspective is conditioned by several constraints or factors, like e.g. costs, technical ability to built-in the solution at particular stage of execution and probability of future use. This open multi-variant approach can often retain some elements of rejected, yet possible in the future further adaptations.

However, at the same time, an element of criticism must be invoked here to refer to the issue of information processing and knowledge processing. Architectural problems are significantly different from engineering problems, and in result contain much larger share of subjective or intersubjective stimuli. This in turn leads to higher instability, higher risk of failures – whether partial or total. Also, both groups of social and cultural factors are prone to the quality of interpretation – causing the risk of misguidance in creating relevant managerial mechanisms. Also, many researchers point out that human component or human factor, to be appropriately respected, should encourage direct participation of people, users of architecture, in the process contrary to usual use of, for example, IT tools supplementing the behaviors of various elements of the system [21, 22]. People add unparalleled unpredictability and creativity both as contributors and decision-making agents, despite certain amount of flaws or errors that may be generated as a result of merging the digital and the analog (or social).

4 COMPLEXITY IN ARCHITECTURAL PROJECT

Quite often theory provides interesting paths to explore, but practice not necessarily recognizes the opportunity to apply it. And architectural practice is the optimal way to prove or disprove the abovementioned theoretical assumptions. While this paper discusses rather the theoretical framework and does not intend to formulate the thesis and advocate for it, one of design applications may be used as partial reference to extend the discourse on complexity into several practical issues used to explore universal nature of application of complexity in architectural design process.

The project of Oborniki Cultural Center (OOK) provided a unique opportunity to perform architectural design, in which there were several elements justifying the implementation of complexity. First of all, the need for new cultural center, built on existing, yet very modest components, was leaving the ambiguity space as to what this new center should become, how it should be programmed, and to which ideas it should refer in order to develop bond between people (future users) and conceived architecture on the one hand, and on the other hand to have a flexible functional solution, rich, but not excessive aesthetic proposal. Simultaneously approaching physical structure and technical requirements multiple issues derivative from the former, confronted with economic and organization related constraints. This represents two levels of engaging complexity – conceptual and definite, latter simply adds various elements to the process (conceptual part remains open for corrections or changes).

The center is planned to serve the community of 35.000 people in 2020 and around 45.000 inhabitants in 2050. The project refers to the old cultural center with current seat in two buildings complex, larger one with main multi-purpose hall in standard relevant to 1980s, and smaller with upper level with insufficient height. Approximate net surface included in the program is 7000 m^2. Project is aimed at delivering the ultimate yet flexible proposal for the development of the center, providing multi-purpose hall, auditorium with relatively large stage and backstage. New building and the arrangement of its surroundings should provide space for social integration, and contextually it should become the main public node for southern part of the town of Oborniki. This particular case is very appropriate because its initial definition is very vague – in fact neither local authorities, nor local community know what ultimately should be the result of the project. There is only an approximation of programmatic assumptions, but the program remains open to any suggestions. There is no ideological content which anyone would like to absorb in architectural form of the building. There are strongly limited resources and financial support, but nobody ever before conducted an analysis on connection between program and investment and maintenance costs.

The project requires the formation of socially binding idea, establishment of the program, functional definition in both terms – architectural and urban. This abstract or disciplinary range of problems is furthermore supplemented with more direct, engineering, technical, or organization-oriented questions, making a network of various factors or agents. While detailed analysis of the structure of design process for OOK goes beyond the scope of this paper, it is still worth noticing that mentioned network corresponds to multiple semi-independent nodes relevant to specific problems or disciplinary issues. For example, economic aspect is depicted by two major groups – expenses and revenues, on lower level expenses refer to physical elements (e.g. shell, loadbearing structure, materials and installations), to media (e.g. energy and water), to manpower or required employment, and revenues, apart from predictable savings, rental revenues, include calculation related to identity gains (brand), social capital gains, etc. One of multiple nodes assumes both internal and external dynamic relationships and shifting – with evolving budget assumptions for particular architectural configuration, which adjusted, imposes reconfiguration among those elements, that are activated in selected setting. This type of organization facilitates otherwise complicated decisions, and this reductive ability does not hamper precision, relevance, and grasp of holistic approach in detailed activity. This process of facilitation relies on exchanging stipulated holistic analysis of the course of the project with series of isolated or connected, extracted little contributions – their impact usually of a small scale, with opportunity to further alterations to seemingly fixed decision. Every content of the project structure was substituted by the factor (or the agent) and the relationships between factors (agents) have been defined as a simple catalogue filled and justified thematically and contextually, but simplified as output for the system. Every factor has stable component and dynamic component, and it may switch its status (from stable to dynamic and vice versa). These interactions could be investigated to find out how complexity can be managed in OOK case study, as shown in connection to Table 2 and Table 3.

Table 2: Catalogue of possible interactions between factors / agents:

Factor / agent A operates at or gets the same level as factor / agent B (has similar impact on decision-making)	A ↔ B
Factor / agent A operates at or gets higher level as factor / agent B and thus A has precedence (has higher impact on decision-making)	A → B
Factor / agent A operates at or gets lower level as factor / agent B and thus B has precedence (has higher impact on decision-making)	A ← B
Factor / agent A operates independently from factor / agent B and thus A and B have no interconnection (difference on impact on decision-making cannot be determined)	A ◯ B
Factors / agents A and B are simultaneously altered by external driver	A ≈ B
One of factors / agents A or B is altered by external driver	A ∞ B
Factor / agent A alters factor / agent B status	A > B
Factor / agent B alters factor / agent A status	A < B
Factors / agents A and B simultaneously alter each other	A ◇ B
Factors / agents A and B are or become (temporarily) disconnected	A X B

Table 3. Factors / agents primary and secondary interactions within design process.

Complexity criteria	Primary interactions	Secondary interactions
Connectivity	A ↔ B	A > B
	A → B	A < B
	A ← B	
Autonomy	A ◯ B	A ∞ B
	A X B	A ◇ B
Emergence	A ◯ B	A ◇ B
	A ≈ B	
	A ∞ B	
Non-equilibrium	A ≈ B	A > B
	A ∞ B	A < B
		A X B
Nonlinearity	A ≈ B	A ◇ B
	A > B	
	A < B	
	A X B	
Self-organization	A ↔ B	A ◯ B
	A → B	A ≈ B
	A ← B	A ∞ B
	A > B	
	A < B	
	A ◇ B	
	A X B	
Co-evolution	A ≈ B	A ↔ B
	A > B	A → B
	A < B	A ← B
	A ◇ B	
	A X B	

Another example of complexity management was the establishment of program of new OOK. On the one hand it required expert driven approach and expert knowledge to propose and to formulate the program. On the other hand it could not be acquired without the participation of current staff of the cultural center, and what's more without acknowledging the needs of amateur bands contributing to the cultural output of the community, having aspirations of being able to permanently use the final OOK architectural complex. *Vox populi* was also necessary to be included, mainly for knowledge generation and socio-political reasons – to enable extension to financial support for this development [23]. This pointed out towards participatory design elements, and ultimately strongly influenced the decision on splitting the task on four different stages – A, B, C and D (Figs. 1 & 2). This contribution was validated

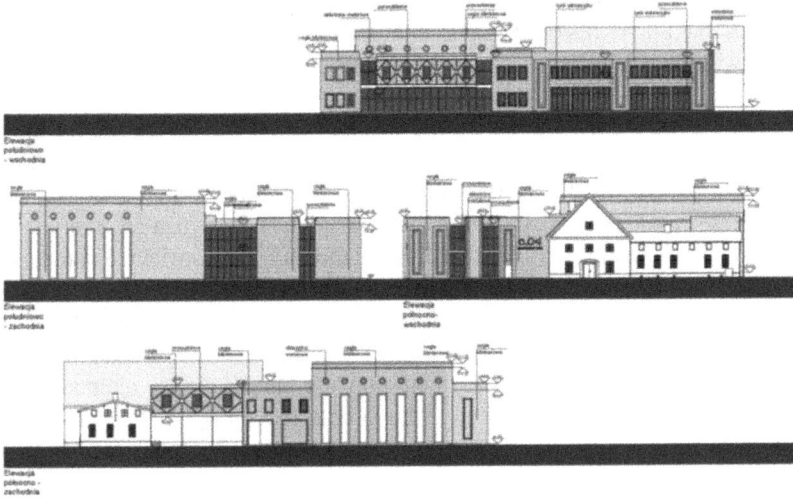

Figure 1: Stage A of the OOK project – facades.

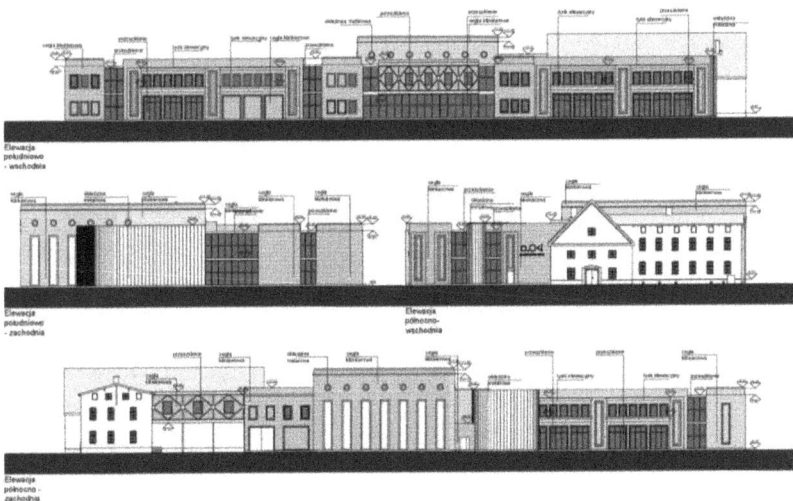

Figure 2: Stage A+B+C+D of the OOK project – facades.

by over 600 respondents, voting for both conceptual solutions and programmatic solutions, and independent elaboration – report – has been produced as a result [24].

The project has been conducted while performing fragmentary participation procedures which lasted for over one year, almost as long as the design itself [25, 26]. It is hardly possible to include community in all key decisions, and hard to organize smooth coordination, therefore design of the center had to be processed and changes induced by potential participatory or external influential factors taken into account during the advancement of defining the architectural shape of new OOK. According to the administration of the cultural center (currently operating) the temporary importance chart for programmatic contents presented itself as (in order of importance): multifunctional auditorium to host multiple artistic events and accommodate artistic activity of multiple bands performing in Oborniki, multifunctional hall, and service areas for performers. Architects' own recognition of needs of community and local bands, but the results of the report exposed the necessity to adjust the programming of the center with the inclusion of a cinema, which in turn, due to some related feasibility analysis launched in relevance, forced the adaptation of multifunctional auditorium to adopt the possibility to be temporary converted into cinema as well [27]. One specialized cinema and two auditoriums with projection rooms added and appropriate equipment granted (Fig. 3).

The project was performed with 6 preliminary organization and architectural concepts, proposing diverse typologies for new OOK (cultural center). Two parallel, evolved concepts were used as potential vessels for spatial and visual identity (with attributed names of Quartz and Showroom, with the latter adopted for final elaboration [Fig. 4]). Even final concept, selected with participation of local authorities and wide group of inhabitants of Oborniki administrative area, was several times altered due to various factors – economic, related to limitation in budget of the development, programmatic, due to will to maintain the overall spatial disposition, or formal derived from cultural code recognized by local community and available in some places of Oborniki as part of its architectural heritage – and OOK as associated volume.

Table 4 exposes selected elements of complexity management within the design process. These are fragmentary and their description due to attributes of complexity cannot be fully explained verbally while not only this implies design process structure vast and

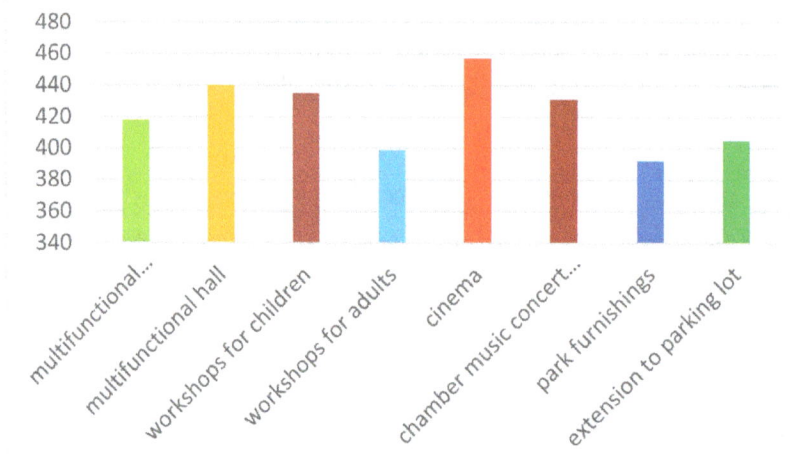

Figure 3: Results of community members' voting for programmatic bias (first 8 positions).

Table 4. Examples of complexity application in OOK project.

Complexity criteria	Architectural design process (examples)
Connectivity	**Intradisciplinary and interdisciplinary networking** roof and gallery definition affecting installation design (HVAC and lighting) and vice versa, energy balance affecting roof structure, form, and skylights distribution technical section is designed in anticipation of future stages to be performed separately, which includes e.g. installation transfers
Autonomy	**Ability to accommodate exclusive** specialized cinema hall was designed as separate, autonomous section of cultural center, with distinct material selection, to maintain future possibility to include or exclude cinema after it was ruled out that stage D will be joint with stage A recording room section retain autonomy throughout the entire process
Emergence	**Inclusion (responsive) mechanisms** inclusion of summer stage design (required) the project initiates social interest, reaction, and expectations, feeding the process 'recycling of material' thread (interior identity anchor)
Non-equilibrium	**Permanent fluctuation (flexible structure)** permanent multiple factors management through the course of design
Nonlinearity	**Open architecture of the process** programmatic adjustments (impact of participatory design content) organization of development stage implementations (1 stage to 3 stage to 4 stages ultimately)
Self-organization	**Permanent reconfiguration (flexible structure)** multiple re-orientation of design course shifts in temporary established hierarchies splitting design process into stages and parallel elaboration of selected elements (e.g. changes of setting of particular contents of the center, obscured facades design)
Co-evolution	**Permanent evolution of contents of the process (flexible structure)** responsive modules within the process anticipation of technologies to be replaced in future programming the process of wearing of architectural substance (use cycles)

multidirectional network, but constant flux of its components. Some connections between factors disappear, others emerge.

Complexity is merged with Meta-Design methodology, which has been frequently applied since 2007, its formulation, analysis, and primary implementations performed in several public sector and private sector commissions. Meta-Design (M-D) connects well with complexity theoretical assumptions. First of all M-D acknowledges and accommodates fundamental shifts in design criteria in its 'meta' project timeline [28]. It includes the idea redefinition option, it retains factors like idea, program, and even the theme (topics) as open ones. It encourages simultaneous autonomization and increased connectivity between contents of the process. While this paper is not intended to describe M-D, its reference serves here the purpose of proving successful implementation of complexity principles in already established and successful design methodology.

Figure 4: Superficial form of OOK project upon completion of current stage (overall design for 4 stages).

5 CONCLUSION

Based on current implementations it is impossible to understand how far complex system may become the main framework of architectural design process, but there is no doubt that the system may and should be embedded into the process to reflect the complexity of the real which is to be depicted in the architectural works. The limits on complex system application is mostly related to human related factors, those same factors that dictate the response of the methodological framework of design process so well facilitated by the complex system elements. This topic has been discussed by Alexander himself [29] advocating for presence of subjective, worldview oriented and inherent bias, which in the theory of science is proven to be inextricable and unavoidable. The observation exposes specific architectural character of project which is rooted in axiology, and connected to human evolution integrated into the project principles. As Alexander puts it in a straightforward manner – generative / evolving architecture is always about living structures, which can be understood as living spaces transformed by people, and is always saturated with ethical (or moral) content [30].

Whether limitations of the system can be overcome remains to be seen, but definitely complexity proves its applicability and relevance in architectural implementations, and responds to what Halina Dunin-Woyseth and Fredrik Nilsson [31] describe as new focus of contemporary progressive practices which is the reflection of social complexity in modern architecture performed in multifaceted urban environments, resulting in going beyond what was traditionally claimed to be architecture. But this innovation still is sinful, still lingers in formal superficial layers of architectural information, so aptly named by Mehaffy and Salingaros architectural myopia. As they elaborately conclude in their considerations, we all:

> (…) must reform the architecture (…-…) without further delay, and place a new emphasis on design that is evidence-based, that pays attention to postoccupancy evaluations, and that, in short, values the outcome for human beings and takes their needs seriously. [32]

and this is the area that application of complexity, at least in design processes already performed by research team, may excel at.

REFERENCES

[1] Buckminster Fuller, R., *Inventory of world resources. human trends and needs*. Southern Illinois University, Carbondale, 1963.
[2] Bürdek, B.E., *Design-Theorie*. Methodische und systematische Verfahren im Industrial Design, Ulm, 1971.
[3] Asimov, M., *Introduction to design*, Prentice Hall, Englewood Cliffs, 1962.
[4] Krick, E.V., *An introduction to engineering and engineering design*. John Wiley and Sons, New York, 1969.
[5] Zeisel, J., *Inquiry by design*. W. W. Norton & Company, New York, 2006. See 27–29. (First edition was published in 1981).
[6] Alexander, C., New concepts in complexity theory arising from studies in architecture: an overview of the four books of the nature of order with emphasis on the scientific problems which are raised. *Katarxis*, **3**, p. 24, 2003, available at: www.katarxis3.com/. (accessed 10 March 2017).
[7] Shoshkes, E., *The design process. case studies in project development*. Whitney Library of Design, New York, 1989.
[8] Farrelly, L., *The fundamentals of architecture*. AVA Publishing, Lausanne, p. 160, 2007.
[9] Rowe, P.G., *Design thinking*. The MIT Press, Cambridge, 1995.
[10] Handler, A.B., *Systems approach to architecture*. American Elsevier Publishing Company, New York, 1970.
[11] Lang, J., *Creating architectural theory: the role of the behavioral sciences in environmental design*. Van Nostrand Reinhold Co., New York, 1987.
[12] Picon, A., Continuity, complexity and emergence: what is real for digital designers? *Perspecta*, **42**, pp. 147–157, 2010.
[13] Ibid.: pp. 156–157.
[14] Op. cit., Alexander, 2003: 9.
[15] Mehaffy, M. & Salingaros, N. A., Architectural myopia: designing for industry, not people. *On the Commons*, pp. 10, 5 and 9.
[16] Rzevski, G. & Skobelev, P., *Managing complexity*, WIT Press, Southampton, 2014.
[17] Upitis, R., School architecture and complexity. *Complicity: An International Journal of Complexity and Education*, **1**(1), pp. 19–38, 2004.
[18] Bektaş, E., *Knowledge sharing strategies for large complex building projects, architecture and the built environment* (No. 04). Delft University of Technology, Delft, The Netherlands. See 139–140, 147–150.
[19] Op. cit., Rzevski and Skobelev, pp. 30–31, 2014.
[20] Barelkowski, R., *The laboratory of theory–practice induction meta-circle. On approaches to architectural design process*. Proceedings from International Conference Impact by designing, Sint Lucas School of Design, Brussels, 2017, April 6th–7th.
[21] Cf. Op. cit., Mehaffy and Salingaros, p. 9, 2011.
[22] Escobar, A., *Notes on the ontology of design*. University of North Carolina, Chapel Hill, 2012. See 41–43.
[23] Anderson, N., Public interest design: expanding architecture and design through process and impact. In J. Cohen-Cruz (ed.), *Hybrid, evolving, and integrative career paths*, Vol. 2, Issue 2, A Journal of Imagining America, pp. 26, 2014. See 12.

[24] Barelkowski, R., Barelkowska, K., Chlasta, L., Rosiak, L. & Wardeski, L., *Opinia lokalnej spolecznosci na temat stanu infrastruktury kulturalnej na terenie Gminy Oborniki i priorytetowych potrzeb w zakresie jej uzupełnienia*. Raport dot. analizy wynikow (Report on results of the research) **2**, September 2016, Armageddon Biuro Projektowe, Poznan – Oborniki, September 2016. See 7.

[25] Barelkowski, R., Barelkowska, K., Chlasta, L., Rosiak, L. & Wardeski, L., *Opinia lokalnej spolecznosci na temat stanu infrastruktury kulturalnej na terenie Gminy Oborniki i priorytetowych potrzeb w zakresie jej uzupełnienia*. Raport dot. analizy wynikow (Report on results of the research), Armageddon Biuro Projektowe, Poznan – Oborniki,.

[26] Op. cit., Barelkowski et al., 2016.

[27] Feasibility exposed that it is financially unreasonable to have one small cinema. Therefore hybrid multiplex with maximum of 3 potential cinema auditoriums has been considered.

[28] Barelkowski, R., Meta-design versus self-contained design. In A. Dutoit, J. Odgers, A. Sharr (eds.), *Quality*, Welsh School of Architecture in Cardiff, Cardiff, pp. 23, 2007.

[29] Op. cit., Alexander, p. 13, 2003.

[30] Ibid.: 16.

[31] Dunin-Woyseth, H. & Nilsson, F., Design education, practice, and research: on building a field of inquiry. *Studies in Material Thinking: Re/materialising Design Education Futures*, **11**, pp. 3–17, 2014. See 8.

[32] Op. cit., Mehaffy and Salingaros, 2011: 9. From the second brackets word "school" was removed, while author believes the idea was to spread this kind of approach to currently operating professional practices, too.

CLASSIFICATION OF TWEETS WITH A MIXED METHOD BASED ON PRAGMATIC CONTENT AND META-INFORMATION

M. ESTEVE, F. MIRÓ & A. RABASA
University Miguel Hernández of Elche, Spain.

ABSTRACT

The sharp rise in social networks in any field of opinion has led to the increasing importance of content analysis. Due to the concretion of the texts published on Twitter from its limitation to 140 characters, this network is the most suitable for the analysis and classification of opinions according to different criteria. Therefore, there are multiple tweet analysis tools oriented from the perspective of semantics for trying to classify content characteristics such as feeling and polarity.

In this paper, the authors present a new approach to classification from a different perspective. The proposed approach addresses a complex mixed model from a perspective of pragmatics, the analysis of opinions in the context of their issuer carried out by a panel of experts, along with the classification of the type of discourse by considering the meta-information of the tweet.

From this new approach, the paper presents a complete and complex analysis process of Big Data, which covers all the characteristic phases of the life cycle: capture, storage, preprocessing and analysis of a tweets database. The aim is to classify the tweets as violent or non-violent in their reference to terrorist acts.

If the classification models based on the metadata of tweets reach acceptable levels of accuracy, this methodology will offer a reliable and semiautomatic alternative for tweet classification.

Keywords: analysis, big data, classification, social networks.

1 INTRODUCTION AND OBJECTIVES

'CRÍMINA' (centre for crime prevention and detection) is working on different fronts related to the prevention and early detection of criminal activities. One of their most current lines of work is oriented towards the classification and search for patterns in the *tweets* posted under different *hashtags* related to terrorist attacks.

Twitter is a huge source of data which can be gathered from opinions freely expressed by users. Five hundred million *tweets* a day are generated [1], which in turn provide 20.5 thousand million data a day to anyone who is able to compile this information. Twitter has a policy to share data freely, and provides APIs which allow a user with developer permits to access its database to compile the *tweets* generated in JSON format. To be exact, it provides API *Streaming*, which returns the *tweets* in real time (*streaming*) according to the consultation being made [2]. For example, the search for three types of *hashtags* (event, humanitarian and *StopIslam*). One of the drawbacks of this API is that the volume of data is 1% of all *tweets* generated in a given time and with the initial bias of the search, which means that the samples obtained from Twitter are representative and biased.

Once a 'clean' enough collection of *tweets* is available, a committee of experts in criminology classifies the content of the *tweets* according to the underlying pragmatics in the main body of the tweet, and supported by a well-defined ontology. Now in the analysis phase, the next step is to generate predictive models of Data Mining which will try to 'learn' the most fitting classification of the *tweets* and replicate it, but this time taking into account the *tweet*'s metadata alone and not its pragmatics. In general lines, this paper aims to present a classification method of *tweets*, starting from learning to classify according to pragmatics and ending by making classifications on the basis of the metadata alone, or the environmental parameters of the *tweet*.

2 STATE OF THE ART

To be able to make accurate enough classifications, it is necessary for the data sets to undergo adequate pre-processing. Although a lot has been published about pre-processing, in this case, as in many others, pre-processing will be a set of specific tasks aimed at cleaning and formatting the data, taking into account that the aim is to perform a classification task about types of discourse.

With respect to the analysis method used when dealing with Big Data from social networks, one of the analytical techniques most in demand is classification according to different criteria of opinions and content that users express on them. Through Data Mining, specially oriented towards extracting valuable information about very extensive data sets, a wide range of descriptive and predictive methods is proposed [3].

To be exact, classification tasks are considered predictive methods which aim to model the different possible outputs from a consequent variable, also called target class variable or dependent variable, based on a set of antecedent variables, also called attributes or independent variables. In the context of classification problems, the consequent variables are required to be nominal, also called categorical, discrete or non-numerical, while the attributes of the antecedent can be both nominal and numerical.

The results of a classification algorithm are sets of rules under the form:

(Attribute1, value); (Attribute2, value); ... → (Consequent, value) (support, confidence)

where support refers to the likelihood of this tuple of the antecedent attribute occurring within the data set, and confidence refers to the conditioned likelihood of the consequent taking this determined value, knowing that the antecedent of this rule has occurred.

Some of the most well-known classification algorithms are ID3 [4] and variations of it, like C4.5 [5] which incorporates a series of improvements to the original algorithm, such as being able to deal directly with numerical attributes (like antecedents) that the algorithm itself segments based on the criteria of gaining information. These types of classification algorithms frequently produce outputs that are not only in the form of rule sets, but also in the form of trees called classification trees, whose interpretation is more immediate.

Nevertheless, the high number of attributes in the antecedent that characterize Big Data problems means that very often classification algorithms are not as accurate and efficient as one would wish. This is because not all the attributes that form the input data set are really important for predicting the consequent, that is to say, not all of them are equally important and many can even become dispensable. In these cases, it becomes necessary to resort to an automatic selection method of characteristics [6]. If a classification algorithm only works with the antecedents that really correlate with the consequent, more accurate and legible models are obtained.

3 INPUT DATA SET

Twitter is a *microblogging* network that permits reading and writing messages on the internet of no more than 140 characters, which are received by anyone who opts to receive them. It is a good source for observing communication because it is used to comment on everything that causes consternation or social interest. Furthermore, the limitation of messages, which on Twitter are no more than 140 characters, allows experts to easily identify the communicative sense of the sender. Based on these limitations, 'CRÍMINA' has defined a basic taxonomy [7] of violent communication from the most basic messages known on Twitter as *tweets*.

The initial method, in this case, is the observation of the phenomenon of violent communication on Twitter as a consequence of criminal acts of terrorism against the population. In order to obtain a sample that would contain a significant amount of references to reactions towards the terrorist attacks two selection criteria were followed. The first was extraction from the three *hashtags* which at some time had been identified as *trending topic* in Spain in the 6 days following the terrorist attack. The second criterion was to try and balance the sample for its analysis from the point of view of the tendencies of the communicative content. So, within the *trending topic hashtags*, labels were selected: one was for humanitarian and supportive contents, another referred to the description of the event, and a third *hashtag* where negative attitudes towards the attackers and their background could be expressed, *#StopIslam*. With these initial criteria, data files with 41 variables each in JSON format were extracted through Twitter API liberated for this purpose, including both original messages and *retweets*.

Below, each *tweet* field is listed:

text, retweet_count, favorited, truncated, id_str, in_reply_to_screen_name, source, retweeted, created_at, in_reply_to_status_id_str, in_reply_to_user_id_str, lang, listed_ count, verified, location, user_id_str, description, geo_enabled, user_created_at, statuses_count, followers_count, favourites_count, protected, user_url, name, time_ zone, user_lang, utc_offset, friends_count, screen_name, country_code, country, place_type, full_name, place_name, place_id, place_lat, place_lon, lat, lon, expended_ url and *url*.

4 COMPLEX SYSTEM METHODOLOGY

4.1 Overview of the analysis process

According to Hand [8] *"Data mining is the analysis of (often) large observational data sets in order to find unsuspected relationships and to summarize the data in novel ways that are both understandable and useful to the data owner"*. The availability of large volumes has generated the need to convert them into useful information and knowledge [9].

In order to analyse and extract something useful from the data, they first need to be available. In some cases, this may seem trivial, being based on a simple data file to be analysed, but in other cases the diversity and size of the sources means that the data compilation process will be a complex task which requires its own methodology and technology [10]. In the case of Twitter, we start from a file in JSON format, which requires a cleaning process before it is analysed. Section 'Compiling *tweets*' outlines the process for compiling *tweets* by making use of API *Streaming* enabled by Twitter.

Data compilation should be accompanied by a cleaning process so that the data are in a condition to be analysed [10]. The benefits of the analysis and of the extraction of knowledge from data depend to a great extent on the quality of the compiled data. Because of the characteristics specific to Data Mining, it is necessary to carry out a transformation of the data to obtain a 'raw material' which suits the exact purpose (for example, discretize the date the *tweet* was created according to its time frame – early morning, morning, afternoon evening

Figure 1: System methodology.

and night), and the techniques that are to be used (automatic selection of characteristics and classification rules).

Once the data are pre-processed, an algorithm for *automatic selection of characteristics* is applied to them, from which we obtain a list of the attributes that most influence the class variable together with their numerical relevance. The class variable is *Do_Dv*, which has been created by the team of inter-judges from 'CRÍMINA' in order to categorize a *tweet* according to whether it is a *tweet* which expresses violent communication or not. In this way, we obtain the metadata of the *tweet* which most influence violent communication, leaving aside the pragmatics of the *tweet*.

Once the metadata of the *tweet* that most influences violent communication have been obtained, the next step is to generate the predictive models of Data Mining which will try to 'learn' the classification of the *tweets* and replicate it, omitting the pragmatics of the *tweets*. These models are represented in the form of a classification tree (Fig. 1).

4.2 Compiling *tweets*

In order to begin analysing and extracting something useful from data, they first need to be available. Twitter is a very attractive platform for investigators for several reasons, one of which is that Twitter is a huge source of data where users can freely express their opinions following the *hashtag* rules, which makes it easy for the investigator to follow the opinions and conversations that are generated on Twitter. The main function of the *hashtag* is to order a large amount of information which is generated on the social networks, allowing users to observe content related to that particular word.

The first step for extracting knowledge from data is to identify and gather the data to be analysed. As we have explained in this paper, the data that we want to extract from Twitter are related to the issues of terrorist attacks that have taken place in Europe in recent years. Therefore, by taking into account the *hashtag* rules that are applied on Twitter, three types of *hashtags* are then selected: event ('*#CharlieHebdo*'), humanitarian ('*#JeSuisCharlie*') and *#StopIslam*.

Twitter offers its users some REST APIs which through programming provide access for users to be able to read and write data from it. The REST API identifies the applications and the users who use *Twitter OAuth*, an industrial protocol for authorization. Because of this, in order to compile data it is necessary to go to the web https://apps.twitter.com/, log in with a user account and create an application which will have the above-mentioned authentication permits *OAuth*.

The next step is to know what our intentions are, that is to say, we should establish whether what we want is to retrieve past *tweets* (maximum 6 days before the current date) or compile the *tweets* that are being generated at that very moment. In our case, when a terrorist attack took place we were interested in compiling the *tweets* that were being generated at that very moment. For this reason, the API that was most suited to us is *API Streaming*.

User	HTTP server	Twitter

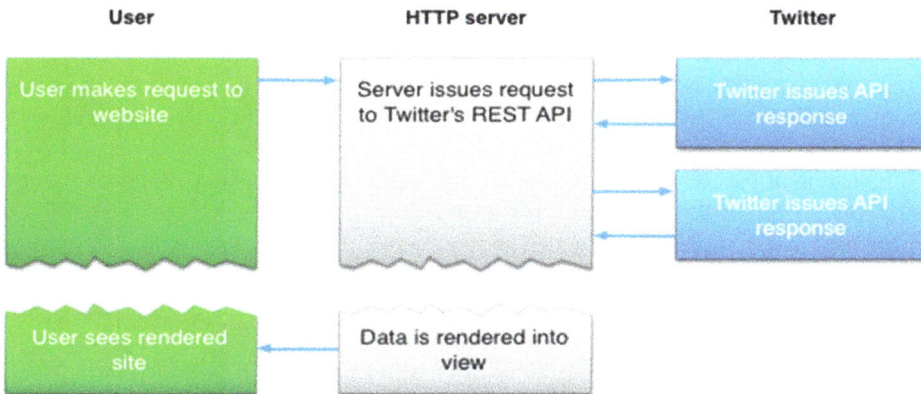

Figure 2: Twitter Apps [11], https://apps.twitter.com/

Figure 2 shows the internal process that is performed when, through programming, a request is made to obtain the data (with the initial bias of *hashtags*) to Twitter. The data that Twitter provides are 41 variables for each *tweet* collected dump, in files in JSON format.

4.3 Preprocessing

Although there is a large bibliography about pre-processing techniques, it is true that this phase is very different for each problem being [12]. It is also only a slightly automatic process where a series of changes are being made iteratively to the original data set. Below we outline some of the frequent steps for data set treatment directed at acting on attributes (columns) and also tuples (rows).

With respect to the data set attributes, the aggregation of characteristics consists of creating new fields to improve the quality, the visualization, or compressibility of the extracted knowledge. It is often necessary to resort to discretization of numerical (or date type) variables to establish ranges or segments that will facilitate the execution of some analytical models that are unable to work with continuous numerical variables. Operations like the elimination of attributes are frequently carried out because they accumulate a dispersion of values that are untreatable in practice. For example, the main body of a *tweet*, or for being primary keys or candidates that identify the registers unequivocally, like a user name. Those attributes, which as a result of some type of selection or bias present for the most part a single value (over 90% of the sample) are also usually eliminated because for practical reasons they are considered uni-valuated. A particular case can be found in the data sets with attributes with a high percentage of null values. Normally, if there is no sense in treating a null value as just another attribute value, and if the concentration of null values is more than 25–30%, this attribute is usually eliminated too. In special cases, where the variables follow a type gradient or known distribution, the null values can be replaced by interpolation. For example, absent values in a series of temperatures.

With regard to the tuples (rows) of the data set, it is necessary to eliminate those that present null values or clearly erroneous values in attributes which are potentially critical for the analysis. For example, a user's date of creation which is post the current date. Also, those registers with a high percentage of null values (above 24–30%) among all their attributes should be eliminated. On the other hand, the values outside the range (outliers) can be treated as null values following the strategy of being replaced when possible or suitable or by eliminating their corresponding row.

To be exact, in the case that concerns us in this paper, the pre-processing actions carried out are the following:

- Elimination of the following attributes because of the high dispersion of values and text: *text, source, Name, Screen name, description* and *Agent.*
- Elimination of user identification attributes and *tweet*: *IdUser.*
- Elimination of attributes with a high value of null values (higher than 70%): *location, url, Time_zone, Geo_Enabled, Coordinate, Entities, Urls, In_reply_to_screen_name, Geo_Lon* and *Geo_Lat.*
- Elimination of the following attributes after being used for the creation of attributes derived from a greater analytical interest: *create_twt* and *create_user.*
- Creation of the attribute: *hoursAnt_Twt, month_usr, TextLength* and *Do_Dv.*

The new attributes created (to be included into the original attribute list), and the possible values they can have are as follows: *hoursAt_Twt* (hours that have passed since the terrorist attack and the creation of the *tweet*), month_usr (months that have passed since the creation of the user and the terrorist attack) and *TextLength* (length in characters of the *tweet*).

4.4 Classification method of the *tweets*

4.4.1 Preliminary analysis
The computational experience described below has been carried out on databases of *tweets* previously classified by experts. These databases are related to the attacks of Charlie Hebdo and Brussels.

Both experiments have seen very similar behaviours in three fundamentals of the study:

- The importance of balancing the sample so that input subsamples are chosen with a ratio of approximately 60% neutral *tweets* and 40% *tweets* of violent communication.
- The similarity of the predictions reached and their respective confusion matrices. In both cases, an average precision of 70% was exceeded, with a higher precision in the classification of neutral *tweets* and a higher false-positive rate.
- The three variables that are extracted as more significant in the face of the prediction of the type of discourse are repeated, albeit in different order.

To simplify the writing and to facilitate the understanding of the work, only the Charlie Hebdo set reclassification is shown.

After the *automatic selection of characteristics* was done, the confusion matrices have been obtained on the original database. Very high mean accuracies were also obtained (close to 98%). However, there was a clear example of over fitting produced by the very high proportion of neutral *tweets* (99.2%) against violent communication *tweets* (0.8%). So, the classification system presents a high number of false negatives that would make it inapplicable in real police contexts.

Thus, the sample of *tweets* was filtered, considering a hypothetical scenario of much social tension, which collected two-thirds of it with neutral content *tweets* and the remaining third with *tweets* of violent communication. The following subsection shows the complete analysis process on said sample.

4.4.2 Classification of the *tweets* as neutral communication (0) or violent communication (1). Charlie Hebdo case.

In this section, *tweets* classified as 0 (neutral) or 1 (violent communication) will be analysed, taking into account the following phases:

1. Selection of more influential attributes on the class variable (with *RandomForestClassifier* algorithms)
2. Confusion matrices and accuracy achieved (with *RandomForestClassifier* algorithms)
3. Classification trees (with *DecissionTreeClassifier* algorithms)

Next, the phases are described:

1. Selection of more influential attributes on the class variable.
   The *RandomForestClassifier* method is used first, to extract a ranking of the attributes most highly correlated with the class variable, as shown in Figure 3.
2. Confusion matrices and accuracy achieved.
   Again, *RandomForestClassifier* is used, but this time to extract the general accuracy and confusion matrix of the classifier, as shown in Fig. 4.

```
                  feature   v_importance
8              TextoLength       0.242672
0              hoursAt_Twt       0.201431
5            statuses_count      0.114815
6                month_usr       0.092647
2            friends_count       0.090591
4          favourites_count      0.088896
1           followers_count      0.088064
3             listed_count       0.069195
7              Geo_Enabled       0.011690
duracion: 0:00:05.922788
```

Figure 3: Most influential attributes.

```
              precision   recall  f1-score   support

          0        0.81     0.86      0.84       804
          1        0.83     0.77      0.80       710

avg / total        0.82     0.82      0.82      1514

PREDICCION    0     1
REAL
0           694   110
1           162   548
```

Figure 4: Confusion matrix and accuracy (*precision*).

The classifier has an average accuracy of 82%. Analysing the confusion matrix, it can be concluded that over the 1,514 instances used by the model, 804 (694 + 110) were neutral tweets ('0'). Of these, 86% (694) were well classified and only the remaining 14% would correspond with false positives (prediction of violent communication, which they really were not). In the case of violent communication ('1'), 710 cases were counted, of which 77% (548) were correctly classified.

3. Classification trees

Using different parameters of the *DecissionTreeClassifier* (Python) model, this section shows and interprets one of the classification trees obtained.

The trees shown use the nomenclature of variables, as follows:

X[0] = hoursAt_Twt	X[4] = favorites_count	X[7] = Geo_Enabled
X[1] = followers_count	X[5] = statuses_count	X[8] = TextLength
X[2] = friends_count	X[6] = month_usr	X[9] = Do_Dv
X[3] = listed_count		

A split binary, three-depth classification tree has been generated with.

The tree represented in Fig. 5 can be interpreted as follows (see the shaded branch): the variable that discriminates most is *TextLength* (*tweet* length in characters) rather than when it is less than 139 characters (≤139.5, i.e. including extra information recently allowed as URLs or links to images). Next, the model finds that the next most discriminating variable is *hoursAt_Twt* (hours elapsed between attack and *tweet*). Specifically, 45.5 hr is a critical threshold value, such that if the elapsed time is less than said value, the tree ends in a leaf node that collects 525 instances: none neutral (and this is very significant). The 525 instances are of violent communication.

4.5 Interpretation of more relevant results

As Figure 5 shows, and looking for the leaves of minimum entropy, the most relevant rules that can be inferred are the following:

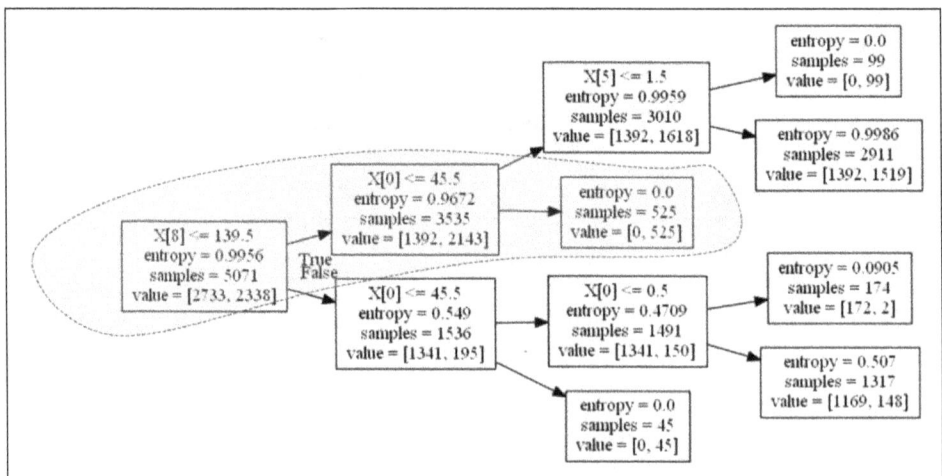

Figure 5: Classification tree.

- If the text does not include URLs or links, the time between the attack and the *tweet* are less than 45 hr and the user has already written a *tweet*, then all samples should be classified as violent communication.
- If the *tweet* incorporates URLs or links and less than an hour has elapsed between the attack and the *tweet*, then only 1.2% of the sample will be classified as violent communication.
- If the *tweet* incorporates URLs or links and it is broadcast more than 45 hr after the attack, then all samples should be classified as violent content.

5 CONCLUSIONS AND FUTURE RESEARCH

5.1 Overcoming the limitations of the language itself

It should be assumed that the pragmatic classification carried out by the experts is based on *tweets* posted in Spanish. This type of classification is dependent on the language the *tweet* is written in and to some extent it is mostly for the expert to judge the ambiguities, turns of speech and context of language. This inevitable point of subjectivity, dependence on language, disappears as soon as the classification of the *tweet* is based only on the environmental issues [13] and not on its semantic and pragmatic content [14].

5.2 Levels of accuracy reached based only on environmental information and not on pragmatics

These levels of accuracy in the classification cannot be interpreted as although the method followed is going to classify any data base of *tweets* with the same accuracy. However, it should be concluded that the model developed is able to classify the *tweets* in almost exactly the same way as the group of experts does and that in some way the algorithm is able to faithfully replicate the criteria of the pragmatic classification in a more objective model of environmental classification based on the meta information of the *tweet*.

On the other hand, as is normal, the accuracy of the classifiers decreases considerably as variables found to have little influence on the selection of characteristics are incorporated. In some scenarios, the accuracy decreases by 40%.

5.3 Possible police uses

The search for hateful or violent discourse potentially generated from criminal situations can become an intractable task to complete within short periods of time, as is evident in the number of *tweets* gathered in the time after a terrorist attack. However, the method proposed facilitates to focus attention on the most relevant variables in each case, that is to say those that have most influence when determining the final nature of the *tweet*. Likewise, and focusing attention on these variables, patterns of potentially conflictive *tweets* can be extracted with high accuracy in much shorter times. In this sense, the proposed system can act both as a filter when searching for opinions which are 'dangerous' for a particular group, and as a system of early alarms when recording a *tweet* whose pattern fits some type of undesirable discourse.

5.4 Other fields of application in crime prevention (radicalization)

Hateful or violent discourse is just another manifestation of the many feelings of users. 'CRÍMINA' is currently working on the classification of *tweets* from the perspective of radicalization, with an immediate practical focus on protecting the European population.

5.5 Other technologies to be incorporated: data streaming

The process presented in this paper corresponds to two completely different sequential phases: first collecting the *tweets* and afterwards their pre-processing and classification. A possible improvement could be to study the viability and suitability of incorporating data streaming technologies which would make it possible to carry out classification processes as data are being compiled. This improvement, which would probably entail a considerable reduction in computation times, is not a small matter at all. It would require automatizing pre-processing phase and pragmatic classification by experts and also coordinating it with the dynamic classification of the *tweets* based on their metadata.

5.6 The importance of adequate pre-processing

It is clear how important adequate pre-processing of data is, and how different pre-processing inevitably lead to different accuracies of the classification models. For this reason, another future line of research to be considered, and one that is absolutely critical, is the systematic analysis of the pre-processing phase and how to implement it, so that the highest possible accuracies can be reached in classification tasks of *tweets*.

5.7 Sub-samples selected to avoid over fitting.

In the two databases studied, it was verified that if a very unbalanced sample is considered regarding the amount of possible values in the class variable, the classifiers also provided severe cases of over fitting, with respect to the class variable. In a new research line, this fact must be formally checked on semi-synthetic sub-samples and under different loading conditions.

ACKNOWLEDGEMENTS
The present paper was carried out in the framework of research project DER2014-53449-R entitled "*Incitación a la violencia y discurso del odio en Internet. Alcance real del fenómeno, tipologías, factores ambientales y límites de la intervención jurídica frente al mismo*", from MINECO.

REFERENCES
[1] Internet Live Stats, Online http://www.internetlivestats.com/twitter-statistics/. (accessed March 2017).
[2] Morstatter, F., Pfeffer, J. & Liu, H., *When is it biased? Assessing the Representativeness of Twitter's Streaming API*. Ed. Cornel University Library, 2014.
[3] Rabasa, A., *Método para la reducción de Sistemas de Reglas de Clasificación por dominios de significancia* (doctoral thesis). University Miguel Hernández of Elche, 2009.

[4] Quinlan, J.R., *Discovering rules by induction from large collections of examples. In D. Michie* (Ed.), Expert systems in the micro electronic age. Edinburgh University Press, 1979.

[5] Quinlan, J.R., *Bagging, Boosting, And C4.5.*, University of Sydney. Technical Report, 2006.

[6] Dernoncourt, D., Hanczar, B. & Zuckera, J.D., Analysis of feature selection stability on high dimension and small sample data. *Computational Statistics and Data Analysis,* **71**, pp. 681–693, 2014.

[7] Miró, F., Taxonomía de la comunicación violenta y el discurso del odio en Internet. *Journal of law and political science studies*, **22**(I), 2016.

[8] Hand, D., Mannila, H. & Smyth, P., *Principles of Data Mining*, Cambridge, MA: The MIT Press, 2001.

[9] Han, J. & Kamber, M., *Data Mining: Concepts and Techniques* (3th. ed.), San Francisco: Morgan Kaufmann, 2012.

[10] Hernández, J., Ramírez, M.J. & Ferri, C., *Introducción a la minería de Datos*, Pearson. Prentice Hall, pp. 19–45, 2004.

[11] Twitter Apps, online https://apps.twitter.com/. Accessed on March 2017.

[12] Wasilewska, A. & Menasalvas, E., *Data Preprocessing and Data Mining as Generalization. Data Mining: Foundations and Practice, 118 of the series Studies in Computational Intelligence*, pp. 469–484, 2008.

[13] Miró, F. & Johnson, S., Cybercrime and Place: Appliying Environmental Criminology to Crimes in Cyberspace. In G. Bruinsma & S. Johnson (eds), *The Oxford Handbook of Environmental Criminology*. Oxford: Oxford University Press, 2017.

[14] Burnap, P. & Williams, M.L., Cyber hate speech on twitter: An application of machine classification and statistical modeling for policy and decision making. *Policy & Internet*, **7**(2), pp. 223–242, 2015.

THE REINFORCED ENTERPRISE BUSINESS ARCHITECTURE (REBAR) ONTOLOGY

CHRISTINE A. HOYLAND
Customer 1 Focus, LLC; Norfolk, Virginia, USA.

ABSTRACT

Understanding organizations and their needs for new technology has never been more challenging than in today's high-tech business world. Enterprise managers are required to coordinate with other departmental managers, direct their personnel and solve problems along the way. Communicating new designs to IT for needed applications may not be in the manager's skillset. When the enterprise grows rapidly or tries to compete in new areas, a set of basic diagrams illustrating common workflows may no longer accurately reflect the complex environment. What is needed is a simple method for illustrating the enterprise as a whole, interoperable structure so managers and workers alike can describe their requirements in the unique vocabulary of their industry. REBAR offers a novel approach for using key strategic and operational business documents, written in natural language, as the basis for the formal enterprise ontology. Popular semantic web standards, including RDF, FOAF and DC, provide generic terms already designed to convey the subject–predicate–object structure of natural language in a social structure. The REBAR enterprise ontology extends these existing standards, thus evolving a socio-technical model of the functional organization distilled directly from existing enterprise documents. REBAR captures the essence of the unique enterprise in a graphical application that can be queried and dynamically recombined to illustrate details of complex workplace collaborations. An enterprise ontology should unite all defined departmental functions authorized by executive enterprise managers. Additionally, findings indicate the REBAR ontology has the potential to provide a reusable structure for linking core social business functions of the enterprise to other explicit enterprise knowledge, including policies, procedures, tech manuals, training documents and project metrics. The REBAR methodology offers evidence that the enterprise is more than the sum of its parts, it is the bridge unifying explicit and tacit knowledge during work projects across the entire enterprise.
Keywords: business plans, enterprise, knowledge management, ontology, semantic web, strategic goals, systems engineering.

1 INTRODUCTION

This paper proposes an enterprise ontology solution that can extract enterprise organizational models from the natural language document enterprise owners and executives implement as their strategic and business plan documents. Reinforced Enterprise Business ARchitectures (REBAR) provides formal, reusable models suitable for developing requirements for interoperable systems that support the entire enterprise.

1.1 Background

In 2009, President Obama issued Executive Order 13520 [1] declaring a focus on reducing improper payments government-wide and eliminating waste and fraud in all government programs through better tracking of government spending and more transparency in the administration of government dollars. Federal agencies are encouraged to analyze their existing rules and strive to achieve greater coordination across agencies to simplify and harmonize redundant, inconsistent, or overlapping requirements, thus reducing cost. To this end, the Government Accounting Office (GAO) [2] is taking on, as its responsibility, the task of pointing out the good, the bad and the ugly of federal technology while providing constructive criticism to help agencies dealing with the highest risk technology investments. US Chief

DOI: 10.2495/DNE-V13-N1-71-81

Information Officer recommends expanding the awareness that these initiatives exist. 'So much of what we do in federal IT is in stovepipes and what we need to do more is leverage existing contracts more – sometimes within an agency,' he said.

2 ANALYSIS OF ENTERPRISE GOALS

Program managers would probably choose to skip building enterprise architecture models of their projects and go right into the system development cycle. Many project managers claim they already know what is wrong with the current system and where it needs to be more efficient. However, because technology vendors want to know the fine details of each process, projects can get bogged down in constructing elaborate, proprietary diagrams.

2.1 Model-based systems implementation

We contend that much of the lack of progress in successfully implementing today's model-based systems engineering methods may be explained by a misplaced focus on detailing numerous how-to activity models. This focus on activities ignores the important elements of who, what, when, where and especially why, that shape and constrain all levels of the enterprise framework: developing, maintaining and facilitating the implementation of a sound and integrated information technology architecture for the executive agency; and promoting the effective and efficient design and operation of all major information resources management processes for the executive agency, including improvements to work processes of the executive agency [3].

Managing a federal program is complicated, and requires that all work that is contracted out is to be managed according to the strict guidelines required by Congress in order to administer the US federal budget. Enterprise models are the blueprints that both program managers and contractors alike can use to agree on a specific set of requirements for the system development project.

2.2 Need for a holistic approach

The role of formal enterprise ontology would be to point to models built during the period of the project as the authoritative reference that should be used as a basis for making decisions on all aspects of the programs (i.e. systems and applications) that follow. However, it is clear from past GAO audits, and other metrics regarding IT project failures, the processes currently used to construct models from bottom-up activity modeling cannot support this claim when scrutinized by independent reviewers. As identified by the audits, core artifacts regarding the enterprise strategic and business levels, which support the agency/segment level mission and business outcomes, are not comprehensive. Instead, models focus on applications, networks and security. While all of these functions are necessary, when built as a separate system, the enterprise will have severe integration problems when attempting to implement the project as a whole. Without a proper foundation regarding the agency/segment level mission and business outcomes, the sub-level architectures will continue to result in stove-piped solutions that do not answer the critical needs of the whole enterprise.

3 ONTOLOGY AS AN APPROACH TO UNDERSTANDING THE ENTERPRISE

An ontology grounded in philosophy is defined as a theory of the nature of existence (e.g. Aristotle's ontology offers primitive categories [4], such as substance and quality, which were presumed to account for *All That Is*). Tom Gruber, a computer scientist at Stanford University, presented a paper in 1993 that formally introduced the analogy of *ontology* to the computer science community [5]. Gruber described his concept of *ontology* as a technical term denoting *an artifact that is designed for a purpose, which is to enable the modeling of knowledge about some domain, real or imagined* [4]. It conveys rules about terms and how they should be used. For that reason, and because ontology languages such as Resource Description Framework (RDF) and Extensible Markup Language (XML) are open standards, they offer the promise of non-proprietary, and therefore reusable terms. When ontology is referenced by an agent or application, it is reasonable to expect greater understanding and increased communication [6].

3.1 Ontology formality continuum

There appear to be two core components of any ontology: a vocabulary of terms and some specification of meaning for the terms. On one end of the continuum would be lightweight ontologies, such as terms, thesauri, and ad hoc hierarchies, consisting mainly of terms with little or no specifications. At the other end are explicit semantics and mathematical logic. Semantic Web technologies, such as Friend-of-a-friend (FOAF), Dublin Core (DC), XML and RDF, occupy the arbitrary center of the continuum, as they appear to have the potential to bridge the gap between natural language descriptions and the formality required for auto-mated processing of semantics [7]. For example, the mature W3C standard for FOAF [8] has been recognized as a formal yet light-weight Semantic Web ontology. Since computers have difficulties interpreting the common language of human beings and in using contextual cues to resolve them, efforts to develop modeling languages specific to the industry classification still require additional resolution [9].

3.2 Enterprise architectures as ontology

Enterprise architectures have been regarded as the blueprints used to understand and change large, complex organizations [10, 11]. As the complexities surrounding development of busi-ness information systems, frameworks that conveyed unique facets of enterprise models were necessary as a way to determine equivalent enterprise views. The Zachman [12] framework, as shown in Figure 1, uses tables to categorize the who, what, when, where, why, and how of architecture viewpoints so that systems engineers could facilitate users' decisions on how to manage technology projects. On the ontology continuum, the Zachman framework would be considered lightweight and informal. While it provides a vocabulary for understanding the various facets of the enterprise change management, it lacks a standard methodology of how to model the architecture within the views. The problem this presents to complex business organizations who attempt to construct enterprise architectures is that *they cannot be read by anyone but the people and programs that created them*, therefore causing the same work to be done over and over again [11].

4 THE NEW ONTOLOGY OF DOCUMENTALITY

The philosophical ontology regarding documentality probes ways in which social reality is created and used [13–15]. The thesis regarding documentality finds that words not only con-vey information, they *bring new types of social entities into being*. For example, there is no physical manifestation of a debt. It is only after two or more specific humans make a promise that the entity of debt can exist. Left unwritten, that debt relies on human memory for its existence. Documents evolved through time have provided humans a way to make up for short or inconsistent memory, since *a document is something that is able to endure*

ENTERPRISE ARCHITECTURE – A Framework						
Level/ (Owner)	DATA (What)	FUNCTION (How)	NETWORK (Where)	PEOPLE (Who)	TIME (When)	MOTIVATION (Why)
ENTERPRISE Scope (Strategic Mgrs)	Lists, Maps and Rich Pictures					Mission Vision, Goals, Objectives
ENTERPRISE Dept (Owner)	Semantic Model	Business Process Model	Logical Network	Work Flow Model	Master Schedule	Business Plan
SYSTEM (Designer)	Logical Data Model	Application Model	Distributed System Architecture	Human-Interface Architecture	Processing Structure	Business Rule Model
TECHNOLOGY (Builder)	Physical Data Model	System Design	System Architecture	Presentation Architecture	Control Structure	Rule Design
DETAILED REPRESENTATIONS (Sub-Contractor)	Data Defintion	Program	Network Architecture	Security Architecture	Timing Definition	Rule Specification
FUNCTIONING ENTERPRISE	e.g. Data	e.g. Function	e.g. Network	e.g. Organization	e.g. Schedule	e.g. Strategy

Figure 1: Zachman-style framework.

self-identically through time. Ferraris extended the documentality concept as he described how social objects come into existence [16]. He asserts that because *Object = Recorded Act*, the social object that results from a given social act is characterized by being registered on a piece of paper, in a computer file, or in the heads of persons. Documents do not write themselves or file themselves away. Humans intentionally create them, sign them, and save them for later use. Without the possibility of inscription there would not even be social objects, such as stocks, pensions or mortgages [17]. Ferraris further identifies a grand divide between strong documents and weak documents. Strong documents are defined as the actual records that represent the social act itself. Weak documents record facts regarding the social object, but are secondary derivatives. And it is the creation and implementation of the strong enterprise documents that provide explicit governance over the entire enterprise.

4.1 Examples of strong documents

The United States Department of Defense (DoD) conveys enterprise strategic and operational plans in the form of military doctrine that is written at a level of detail general enough to cover numerous mission variations. Yet this library of information is specific enough to convey the requirements that military leaders have agreed are important. These strong documents define authorized social acts, in the form of required communications with various segments of the organization. Included in doctrine are references to inter-governmental and non-governmental organizations, as well as various Federal organizations involved in mission success. As such, doctrine describes the social entities and their networks, with emphasis on authorized and required communications [18]. When the REBAR extension of the FOAF ontology is applied to a sample of Joint doctrine, the model of a diverse and creative organization begins to takes shape. In this way, REBAR models at the strategic and business-level describe the *minimum critical specification* of what is absolutely required, that is, the strong documents that instantiate the enterprise [19].

With this knowledge specifically linked to users' needs and wants, enterprise project managers will be able to accurately describe their requirements, including *hyperlinks back to the specific paragraph* in the document that illustrates *why* these elements are required. This REBAR model is necessary in order to convey not only the technical standards required for new technology, but also governance standards required by law, policy, rules, regulations and the formally accepted culture of the organization.

5 THE LACK OF FORMALITY GAP

It is not difficult to see how detailed process diagrams became the official artifact used to convey what was going on in our organizations currently. The assumption inferred that if workers identified *how* they accomplish work now, vendors could supply technology that would take them to the next level in productivity. While industries that use automation to mass produce precision parts can see benefit from detailed activity study, the practice becomes problematic when, as Zachman recommends, the analogy of the factory system is extended to the non-routine technical system. Sociotechnical systems (STS) theory was applied to engineering around 1982, providing the shift from emphasis on the activity to what results from the action, i.e. the product [20].

5.1 REBAR, the proposed solution

At the strategic and operational levels of the enterprise, governance documents are written [2] in general language so that the directives may still be relevant in a variety of situations. This level of direction dictates *who* is responsible *when* a triggering event occurs, and *what* should be considered at that time.

When a document is implemented in an enterprise, this action effectively authorizes action to occur. Looking at the whole enterprise document library, it becomes clear that the enterprise is a complex social network, relying on interactivity between teams to solve problems within the boundaries set by these governance documents. In order to have utility when used in a variety of situations, details describing the specific activities and tools that support these activities are left out of doctrine. It remains up to the discretion of the well-trained team tasked with performing the activity to collaborate on *how* they will carry out the plan.

5.2 REBAR, the rapid, authoritative, holistic approach

REBAR is a patent-pending process for tagging governance documents with meta-data so they can objectively illustrate the holistic enterprise at its essential level. REBAR follows the standards established by the World Wide Web Consortium (W3C) [21]. The resultant visualizations go well beyond basic cluster diagrams indicative of social network representations. By capturing interactivity between working groups that include what they produce and the implied subjects regarding their collaborations, dynamic threads of communications can be pulled to represent a wide variety of events.

Unique models that depict specific context surrounding events can be dynamically constructed using the REBAR working model builder. These visualizations show actors collaborating together on their enterprise process. They are able to choose from lists of authorized options that satisfy the requirements of the situation. They can then customize their version of the scenario based on specific contextual criteria. These new scenario threads retain links back to the overarching, unmodified strategic or operational policy, goal, or objective that the project participants are seeking to satisfy. Perpetual linking provides the basis of comparison and a qualitative method for assessing how well the solution achieves the performance goal. The core requirements of the enterprise remain stable and visible throughout the analysis, thus continuing to provide references as justification of the requirement [19].

5.3 Configuration management for strong enterprise documents

Since strong documents at the governance level are maintained using stringent configuration management controls, are regarded as authoritative, and are written to remain in effect for long periods of time, effort required to tag these documents quickly returns a benefit higher than the cost of populating this enterprise knowledge library. The natural language and rich but unstructured pictures of doctrine documents is parsed by personnel trained in using a research technique, namely content analysis. The result was the creation of the interactive, semantic REBAR model methodology. Importantly this model is traceable, that is, each object in the model is linked by a uniform resource locator (URL) back to its original position in doctrine, and can be displayed at the click of a mouse. Functional requirements developed using REBAR references are objective, i.e. they are not influenced by personal opinions, only by authorized written words and rich pictures.

Current enterprise models are built and stored using complex modeling tools. While technically oriented personnel may be able to follow the logic of these models, functional subject matter experts would not be able to review and assess the models produced using these tools. As a result, the models that are supposed to provide the *blueprints* for the future enterprise may not get the reviews and corrections necessary to prevent errors. The REBAR methodology instead offers models built using unique vocabulary of the industry the enterprise already uses on a daily basis. Using the REBAR visual tools for queries expands the utility of the methodology. Enterprise personnel can be given access to techniques for conveying their natural language requirements to their IT providers in a way that makes sense to the most important people on the project, i.e. their sponsors and customers.

6 RESEARCH RESULTS

For this study, the research design directed construction of the REBAR XML schema using W3C Semantic Web standards to capture the essential components of the organization. This schema was then used as a data-coding and collection instrument designed to facilitate content analysis of the selected sample of Joint doctrine. The first-stage analysis culminated in an assessment of the potential of the approach for representing Joint doctrine as a dependable strategic/operational-level enterprise model of Joint force logistics. Using the W3C standards for constructing a schema [22], this process was completed in several iterations.

6.1 Sample application selection

First, the entire collection of Joint doctrine publications in the Joint Electronic Library was examined. Collectively, all publications that are approved for use by Joint Forces would be considered as the holistic description of the mission known as the Department of Defense military decision-making process. The approved Joint Doctrine publication, JP 4-0, *Joint Logistics* was selected as a good sample. It was freely available online, representative of the Library population because it embodies both strategic and operational tenets of Joint doctrine, and includes descriptions of planning, execution and control operations; including those in cooperation with multinational partners and other US Government agencies

6.2 Schema correction and use

The construction of the XML schema was finalized. It was composed of three classes; organization, agent, and document, from the FOAF standard [23] for the initial REBAR schema. Several FOAF predicates that could also describe types of communication between DoD organizations were selected. For example, the statement 'Agent (is a) *member* (of the) Organization' would represent the block of text in Joint doctrine stating the Secretary of Defense (is a) *member* (of the) Office of the Secretary of Defense.

The FOAF standard uses the predicate *knows* to indicate linkage between two organizations. However, when scanning various blocks of text in doctrine, it became clear that communications between organizations were of a more intense nature than mere acknowledgement. Accurately representing Joint doctrine that assigns required tasks or requests urgent information necessitated more descriptive predicates. Also, in some cases, the directionality of communication is described in doctrine, indicating message traffic flow. Therefore, in order to represent the text *as written* in doctrine, several predicates were added to the REBAR schema, including *sendMessageTo* and *receiveMessageFrom.* Having several ways to represent doctrine-implied communication proved useful for replicating authorized directives to accurately convey their meaning. For example, the block of doctrine text that indicates ongoing collaboration between two organizations is represented by the use of both *sendMessageTo* and *receiveMessageFrom* predicates. This dual-predicate use serves to characterize continuous communications when this is what is implied by doctrine.

6.3 Resolving missing elements

While assessing the use of FOAF standards in an early draft of the REBAR schema, the researcher noticed there was no way to communicate *what* the teams were exchanging information about, other than to link to their *workplaceHomepage.* The researcher resolved this dilemma by constructing *message* as a type of Document that could convey topics. Since *message* is a type of document, the attribute *type* was added to the REBAR schema Document class. By adding the attribute *type,* doctrine text that describes a planning document is referred to as *type = document,* while a document that is a message is referenced as a document *type = message.* In the REBAR schema, the predicate *title* that is associated with Document is now coded as the message subject line, while the predicate *description* is coded to signify the word-for-word description referenced by doctrine.

The document type *message* is useful for illustrating two aspects of Zachman's enterprise architecture taxonomy. First, the act of receiving a message can trigger some event, that is, *when* the message is received, something needs to happen. Second, doctrine also describes *what* should be done in response to events, or triggers. These details of doctrine text are conveyed as the message *title* and *description*, directed at Organizations who are required to act *when* these events occur.

6.4 Validating presence of event descriptions

The researcher also noticed that Joint doctrine describes various events using action verbs. In Joint doctrine, the actions are described in general language; i.e. no specifics are used to describe *how* to perform *what* needs to be done. This observation confirmed that the REBAR approach was appropriate for uncovering important triggers and events described at the enterprise strategic- and operational-levels while also providing rules for parsing doctrine text describing actions.

Adding the implied message title and specific message description adds the formality required for more precise search descriptions. For example, when personnel want to search through procedural documents or tech manuals to find the details regarding *how* to accomplish a task, the REBAR message metadata provides more precise search returns from online libraries the enterprise has processed using current rule-based artificial intelligence (AI) technology. REBAR has the potential to eliminate a good portion of the tedious work personnel must perform in order to *train* off-the-shelf learning systems for use with their specific enterprise content.

The amount of text that could be coded during content analysis using the REBAR schema increased substantially when the message *type* was added to the data collection schema. There were several other changes and additions made to the initial version of the REBAR schema based on execution of the research design. The following modifications served to refine it for use as the prototype: xReference – the third *type* of Document class. This attribute represents documents that are cross-referenced in doctrine for more information about the topic described, and swimlane – an attribute that describes the level of hierarchy of an organization.

6.5 Research – in summary

Stage one of the research was designed to provide answers to the research question, that is, *what are the most significant factors to consider when translating authoritative text and rich pictures into semantic models*? To start with, a representative sample of Joint doctrine was selected. Then the REBAR schema, based on the FOAF standard, was developed in iterations. The FOAF standard was extended with several new predicates so that the text *as written* in Joint doctrine could be accurately represented. An important discovery was made while developing the schema. It became apparent that service-provider organizations react to events that trigger certain responses. Therefore a message type of document object was created to handle event descriptions. Once the instructions for defining this message object were added to the parsing instructions, the amount of doctrine text that could be categorized using the REBAR schema increased greatly. The process for parsing sample documents was reevaluated periodically and revised to include more specific details that would make the parsing process less subjective. Also, acceptable criteria for parsing certain Joint doctrine blocks of text and rich pictures, such as how to identify key words as used in tables of contents, overviews and summaries; were added to the parsing instructions.

7 VISUALIZATIONS

Once the completed XML document is saved to the server, it can be made available to authorized users. As shown in Figure 2, the digital library for the enterprise can be viewed as web pages. Queries in the form of logic statements regarding various teams within the parent organization are displayed based on selections the user wants to research. Organizational descriptions, memberships, work products, communications and collaborations are linked together to show specifics regarding both details and references back to the original document block.

Along with interesting profiles of the teams, came the emergence of the knowledge structure of the enterprise. Cross-referenced documents became visible, and access to documents was facilitated by direct links to portable document formant (PDF) so the user could examine the reference on the spot. Document cards and documents chart visualizations started to show

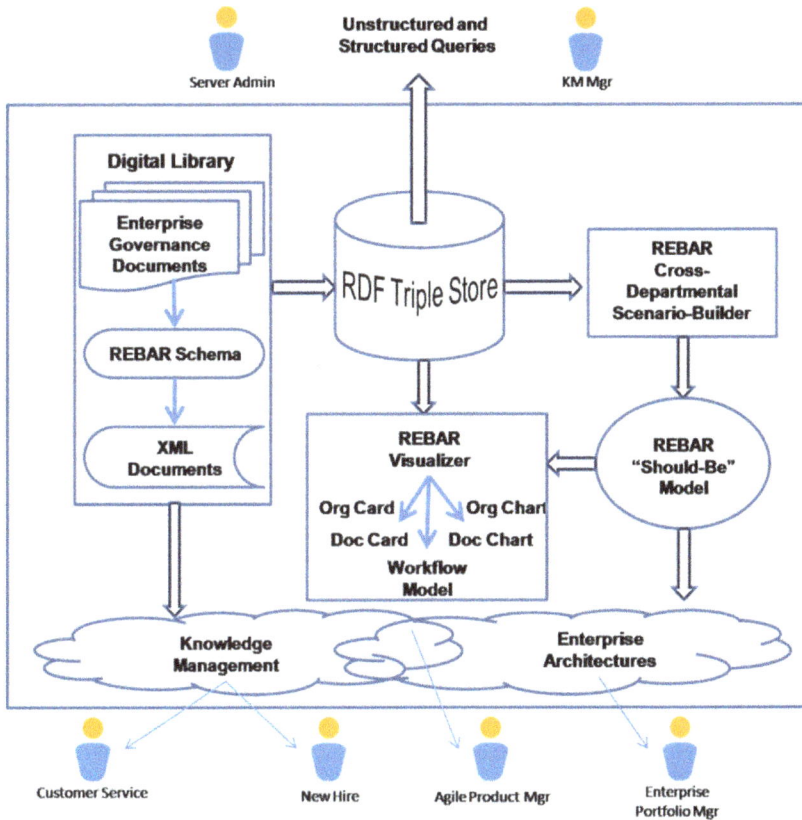

Figure 2: REBAR digital library.

links from the Joint doctrine strong documents to weaker derivative documents, such as Universal Joint Task Lists (UJTLs), Joint Capability Areas (JCA), and Joint Staff policies, plans, procedures, lessons learned and more. Growing the enterprise digital library of documents as linked sets was another happy surprise.

Natural language processors, such as IBM's cognitive AI application Watson, identify methods for *training* an installed instance of the application. This involves humans feeding search terms into the application and then verifying or correcting Watson's response. It would seem that the meta-data provided during the parsing phase of populating a REBAR digital library would provide a good start on the training an application like Watson needs.

8 CONCLUSIONS

Changing the approach from activity-focused models to collaboration and communication visualizations of the enterprise organization unleashes powerful parallels to social network metaphors. Stove-piped solutions that produce islands of technology can be avoided by employing better enterprise change planning. When innovative solutions are sought in answer to complex conditions and an array of standards, strategic and operational governance has proven its value as guidance to its users. It promises no less when used to form the adaptable and flexible ontology of the entire enterprise organization, including valuable links to the many diverse organizations that make up the complex enterprise. The REBAR methodology produces models at the strategic and business-level of the enterprise. As an example, the DoD provides strong documents in the form of military doctrine, DoD policy, procedure, laws, rules, regulations and other documents that it keeps current and makes available to it personnel. Because they are written in natural language, strong documents are difficult to query using intelligent, semantic web tools. The REBAR methodology offers a corresponding formal semantic model that enables users to interact with the mission threads discussed in authoritative documents, and produce dynamic models of the challenges they face as they work to implement that guidance as they carry out their mission assignments. While there are numerous uses for this concept in both military and non-military institutions alike, the REBAR methodology was developed to provide a new means for making sense of complex enterprise organizations. Because of the formality of REBAR, it is possible that derivative documents could be processed by automated applications using the parsed REBAR metadata. And considering that many participants in the enterprise become expert in performing only a unique part of the mission, it is important to always be able to research and understand the organization as a whole.

REFERENCES

[1] Obama, B., Reducing improper payments. In *Federal Register*, Washington, D.C., 2009.

[2] GAO, *Effective Pracctices and Federal Challenges in Applying Agile Methods (GAO-12-681)*, Washington, D.C., 2012.

[3] NASA. (2015, 04-02-2015). FY 2014 Annual Performance Report and FY 2016 Annual Performance Plan, available at http://www.nasa.gov/sites/default/files/files/NASA_FY14_APRFY16_APP_Complete.pdf

[4] Gruber, T., Ontology. In *Encyclopedia of Database Systems*, eds L. Liu & M.T. Özsu, 2008.

[5] DoDI, Interoperability of information technology (IT), Including National Security Systems (NSS). In *DoDI 8330.01*, Washington, DC, 2014.

[6] Uschold, M. & Gruninger, M., Ontologies and semantics for seamless connectivity. *ACIM SIGMod Record*, **33**, pp. 58–64, 2004. https://doi.org/10.1145/1041410.1041420

[7] Gruninger, M., Enterprise modelling. In *Handbook on Enterprise Architecture*, eds P.B. Bernus, L. Nemes & G.J. Schmidt, Springer: New York, pp. 515–541, 2003. https://doi.org/10.1007/978-3-540-24744-9_14

[8] Graves, M., Constabaris, A., Brickley, D. & Miller, L., FOAF: Connecting people on the semantic web. *Cataloging & Classification Quarterly*, **43**, pp. 191–202, 2007. https://doi.org/10.1300/J104v43n03_10

[9] Morosoff, P., Rudnicki, R., Bryant, J., Farrell, R. & Smith, B., Joint doctrine ontology: a benchmark for military information systems interoperability. In *Semantic Technology for Intelligence, Defense and Security (STIDS)*, 1325, pp. 2–9, 2015.

[10] Ross, D.T., Applications and extensions of SADT. *Computer*, **18**, pp. 25–34, 1985. https://doi.org/10.1109/MC.1985.1662862

[11] Wisnosky, D.E., Engineering enterprise architecture: call to action. *Common Defense Quarterly*, pp. 9–14, 2011.

[12] Zachman, J.A., A framework for information systems architecture. *Ibm Systems Journal*, **26**, pp. 276–292, 1987. https://doi.org/10.1147/sj.263.0276

[13] Smith, B., How to do things with documents. *Rivista di Estetica*, **50**, pp. 179–198, 2012.

[14] Ferraris, M., Documentality or why nothing social exists beyond the text. In *Proceedings of the 29th International Ludwig Wittgenstein-Symposium*, Kirchberg, Austria, pp. 385–401, 2007.
https://doi.org/10.1515/9783110328936.385

[15] Searle, J., *The Construction of Social Reality,* vol. 2, Free Press, 1995.

[16] Ferraris, M. & Torrengo, G., Documentality: a theory of social reality. *Rivista di Estetica,* **57**, pp. 11–27, 2014.
https://doi.org/10.4000/estetica.629

[17] Ferraris, M., Collective intentionality or documentality? *Philosophy and Social Criticism,* **41**, pp. 423–433, 2015.
https://doi.org/10.1177/0191453715577741

[18] CJCS, *Interagency Coordination During Joint Operations vol. 1,* ed. J. Staff, Washington, DC, p. 96, 1996.

[19] Hoyland, C.A., *The RQ-Tech Methodology: A New Paradigm for Conceptualizing Strategic Enterprise Architectures.* Doctor of Philosophy Dissertation, Engineering Management, Old Dominion University, Norfolk, VA, 2013.

[20] Taylor, J.C. & Felten, D.F., *Performance by Design: Sociotechnical Systems in North America*, Englewood Cliffs: Prentice-Hall, Inc., 1993.

[21] W3C, Resource Description Framework (RDF) Model and Syntax Specification, ed, 1999.

[22] W3C, XML Schema Part 1: Structures Second Edition, 2004.

[23] Brickley, D. & Miller, L., Introducing FOAF. In *FOAF Project*, 2000.

A COMPLEXITY PERSPECTIVE ON CSR AND SUSTAINABILITY: THEORY AND A LONGITUDINAL CASE STUDY

TERRY B. PORTER & RANDALL REISCHER
The Maine Business School, University of Maine, USA.

ABSTRACT

Corporate social responsibility (CSR) has become standard practice for many if not most of the business organizations today. Extant CSR research largely assumes positivist, linear, and reductionist epistemologies and frequently invokes a 'systems theory' that is unspecified, seemingly taken for granted. The dominance of this conventional approach obscures or conflates significant dynamics that complexity systems theory reveals. This paper develops and compares an emerging complexity theory view of CSR with the more conventional approach. A longitudinal quantitative case study then tests competing hypotheses, permitting an examination of the efficacy of each approach for understanding CSR adoption processes, to our knowledge for the first time.

Two longitudinal surveys of employee attitudes were administered during, and one year after, the rollout of an internal CSR initiative in a Welsh civil society organization. Hypothesis testing led to three anomalous results. First, expected linear increases in employee wellbeing did not persist; second, neither expected nor latent attitude constructs were in evidence; and finally, the only attitude items to increase significantly over time were related to tangible social action and interaction, over and above changes in thinking and beliefs. Complexity theory offers alternative explanations of these results, explanations which we propose be developed to provide a more complete understanding of CSR practice and theory.

Keywords: company responsibility, complexity theory, corporate social responsibility, high performance work systems, modern working practices, resilience, strategic human resources management, sustainability, sustainability adoption processes.

1 INTRODUCTION

Corporate social responsibility (CSR) is becoming ubiquitous in business organizations. Documented benefits to companies include improved financial performance, enhanced legitimacy and image, product quality, operational efficiencies, decreased risk, and attractiveness to investors, among others [1]. For employees, it has been found that perceived intrinsic rewards, trust in managers, decreased stress, job satisfaction, personal wellbeing, and willingness to invest 'discretionary effort' increase when greater attention and effort is put into CSR [2]. With so much good news, managers are inquiring into how to implement an effective CSR program. Yet here the research falls short, for very little has been studied or learned about internal employee-centric processes related to CSR.

Our aim is to examine the internal processes that lie behind sustainability adoption decisions, both theoretically and empirically. Our literature review finds two distinct perspectives to be in play, the taken-for-granted linear systems approach, and an emerging viewpoint based in complexity theory. From here we develop a set of opposing hypotheses, which we then test in a longitudinal study of a sustainability focused human resources initiative in a UK organization. Thus, we have a means of comparing quantitative findings from each perspective, enabling comment on whether and how each approach contributes to CSR praxis.

© 2018 WIT Press, www.witpress.com
DOI: 10.2495/DNE-V13-N1-82-92

For purposes of this study we define CSR as voluntary operational goals and activities that aim to advance a social or environmental agenda beyond the interest of the firm and beyond legal requirements [1, 3]. In contrast to some scholarship, we distinguish sustainability from CSR, as a more macro concept that includes visioning, strategizing, and taking action towards a world with enough access to resources for all, a healthy biosphere, and other broad spectrum ideals [4]. CSR and sustainability overlap of course, and we employ a third term, company responsibility (CR), as an umbrella concept to encompass both.

2 CONCEPTUAL FRAMEWORK

2.1 Conventional research perspective

Several major reviews of the CSR literature concur in finding a small but positive effect of CSR programs on company financial performance (CFP) [1, 5, 6], in addition to the company and employee benefits mentioned above. Yet despite these valuable findings, the responsible mechanisms or linkages are unclear, largely because they involve internal organizational processes that are not well understood [2, 6]. Indeed, Aguinis and Glavas [1] found employee-level perceptions and actions addressed in only 4% of the 690 papers and chapters they reviewed. Further, most studies rely on large, cross-sectional, firm-level datasets that compare antecedents and outcomes [6]. Cross-sectional research, while valuable, is unable to provide insight into the ongoing processes of change and development that underlie organizations' continual search for positive CR results.

A separate issue with this literature is that 'systems theory' is frequently invoked but rarely defined. We presume the reason to be that only one foundational theory of systems is assumed, that based in the positivist, linear, and reductionist assumptions of traditional, neoclassical management theory. There is a search for causality and a drive for a unifying, generalizable theory. For purposes of this paper we label this as the 'conventional' approach in the CSR research.

2.2 Complexity theory perspective

There are dissenting voices in CSR discourse as well, voices critiquing conventional approaches and urging alternative directions. Margolis and Walsh's [7] call to transcend the economic imperative of neoclassical organization theory in CSR research remains a strong beacon today. They recognize that existing management theories 'may be too simplistic and static to fully explain the complexity of the paradoxical demands inherent in the management of sustainability' [8]. Such critiques and calls point toward a radically different approach, that of complexity theory and complex adaptive systems (CAS).

We refer the reader to other, fuller explanations of complexity theory and its application to management and sustainability [9, 10]. The primary premise is that individual and idiosyncratic actions of line level actors are the core and central process of system development. This focus on the individual *as an individual* is precisely what we need to get to the hard-to-reach employee-process level of CR adoption. The problem, and perhaps one reason it has not been much employed to date, is that complexity theory assumptions are incommensurable with those of conventional systems research. The two theories cannot be selectively spliced onto one another. This is because CAS are non-linear, self-organizing, and highly unpredictable. They are continually created in the moment by many types of actors whose directed self-interest brings them into contact with many other actors. The resultant melee is uncontrollable, but regularities do emerge over time through coevolution and emergence [11]. Each system and context is unique, such that generalizing across organizations, geographies, and cultures is not warranted [12]. Henceforth we label this complexity theory perspective as the 'complexity' approach.

2.3 High performance work systems

An established entrée to employee-level social processes is that of Strategic Human Resource Management (SHRM). Beginning in the 1980s, SHRM initiatives, known as High Performance Work Systems (HPWS), Modern Working Practices and other monikers, have been successful in increasing employees' knowledge, skills, empowerment and motivation to contribute 'discretionary effort' [2]. Initiatives involve customized combinations of flatter hierarchies, increased employee participation, flexible work schedules including working from home, redesigned office spaces, reduced supervision, individual control of tasks, and revised incentive systems linked to creativity and productivity [13, 14]. Analyses of HPWS generally concur in finding a clear link with company financial performance (CFP) [15], though not without exceptions [16]. Henceforth we inclusively refer to this type of program as HPWS.

Interestingly, HPWS programs have also been linked with CSR and sustainability through the mediator of Company Social Performance (CSP) [2, 17]. CSP is an overall snapshot of the firm's social posture [18], and includes both internal employee wellbeing and outward-facing service to community. By focusing on employee wellbeing, HPWS is both a significant part of CR and a window to its micro-level adoption. As with CSR research, however, this research stream is limited by its unacknowledged embeddedness within Conventional assumptions. Most studies are cross-sectional [12], lack consensus on measures and methods [19], and provide little insight into the relationships and processes that link employee attitudes with CSP and CFP [20]. For these reasons, a complexity approach using primary data and a longitudinal perspective is called for, to better understand the 'black box' of HPWS employee dynamics [20, 21]. In the empirical study that follows we problematize the unclarified use of 'systems theory,' take a longitudinal perspective and a microprocess level of analysis, and focus on employees as lynchpins between HPWS and CSR intentions and achievement.

3 CASE STUDY AND HYPOTHESES

Our empirical project was a two-year quasi-ethnographic case study of a public sector organization in Wales. As we began, the home office of the organization was rolling out a comprehensive HPWS project as a follow-on to an extensive strategic re-view and re-vision in the previous year. The present study focused on employees' perceptions and responses to the rollout through two surveys administered at three time points: as the rollout was beginning, as the rollout ended six weeks later, and one year after the rollout (see table). Competing hypotheses were developed from conventional and complexity perspectives, which enabled us to compare theoretical predictions in a live setting. The low Ns reflect the small size of the organization, and while they result in low statistical power they also represent very high response rates, 70% or better. Although the surveys were anonymous, the participants at earlier time points continued to participate later. Time 3 was much larger as participation spread beyond the organization's home office.

Table 1: Two surveys and their administration.

Title	Source	Time 1 (t1), May 2014	Time 2 (t2), June 2014	Time 3 (t3), June 2015
Zinger Survey	Single-item feelings of wellbeing	✓	✓	✓
Attitude Survey	Validated attitude scales		✓	✓
		N = 13	N = 16	N = 36

3.1 Hypotheses

3.1.1 Employee wellbeing

The first survey, nicknamed the Zinger, was a simple, quick survey designed to capture employees' spontaneous responses to the HPWS project at the three time points shown above. The survey elicited employees' self-report on single items representing eight feelings in the workplace on a 5-point Likert scale: these were creativity, productivity, collegiality, collaboration, job satisfaction, autonomy, engagement, and positive morale.

Interestingly, prior conventional research has shown an increase in most of these feelings following CSR and HPWS initiatives (see Introduction and Section 2.3). We have no reason to expect a different result here, between t1 and t2. Less researched, but of great interest, is what happens between t2 and t3 of the study. We reasoned that conventional approaches would predict a sustained increase between t2 and t3 as well. In a linear systems framework each of the eight feelings is an independent variable that is stable and predictable over time. It would therefore respond consistently to the same stimulus; in this case, the rollout and incorporation of HPWS changes into company routines.

Complexity theory, on the other hand, does not isolate pairwise relationships and look for longitudinal causality. Indeed, linear causality is a non sequitur in CAS. Instead, the coevolutionary dynamics and continual self-organization of the system means that any one actor could be both influencing and being influenced by another at the same time. Further, feelings may change, dissolve, or evolve in unpredictable ways in CAS. They are social constructions amongst those party to them at the moment [22]. Hence, from a complexity perspective, we do not expect a linear trend in the longitudinal Zinger Survey results. To summarize:

Hypothesis 1a: The three time points measured in the HPWS project will show improvement in a linear relation over time.

Hypothesis 1b: The three time points measured in the HPWS project will not follow a linear trend.

3.1.2 Employee attitudes

The second survey, called the Attitude survey, was more in-depth, designed to study attitudinal changes that have been previously related to HPWS projects and CSR. We extracted four items each from six validated attitude scales (in order to keep the survey to a manageable length), as shown below. Respondents were asked to rate the strength of their agreement with each statement on a 4-point Likert Scale.

Table 2: Attitude survey: constructs and items, (r) denotes reverse scoring.

#	Attitude Construct	Item
1		I have ideas about how to improve this company.
2		People sometimes talk with me about their new ideas and initiatives.
3	Innovation	A group of us are trying out a new idea here.
4		I try to promote new ideas and initiatives to others.
5		I feel that I have a number of good qualities.
6		All in all, I am inclined to feel that I am a failure. (r)
7	Self-esteem	I am able to do things as well as most people.
8		I take a positive attitude towards myself.
9		I am proud to be an employee of this company.
10		In general, this company's goals are similar to my own.
11	Organizational Identification	I find it difficult to agree with this company's policies on matters familiar to me. (r)
12		I find that my values and the values of this company are very similar.
13		I am a person who cares about sustainability.
14		Let future generations solve their own problems. (r)
15	Sustainability Attitude	The positive benefits of economic growth outweigh any negative environmental or social impact. (r)
16		I don't buy consumer products if I know that unethical, unjust, or harmful practices were involved in their production.
17		I feel there is never enough time to get things done. (r)
18	Work Stress	My work role tends to interfere with my personal life. (r)
19		The amount of work I do interferes with the quality I want to maintain. (r)
20		I do not get enough resources to be effective in my job. (r)
21		If I make plans, I generally succeed in executing them.
22		If I have a failure the first time, I keep working at it until it is going better.
23	Self-efficacy	I am usually able to solve problems well in my life.
24		I do not start learning new things if I think they are too difficult. (r)

Since this survey was concerned with multi-item variables, our first hypothesis was a test of their reliability in representing each variable, and their consistency in doing so at both time measures. A complexity perspective would point out that even this simple test belies conventional assumptions without acknowledgement as such. In complex systems, attitudes are socially constructed in dynamic local contexts and likely to change as relationships and contexts change over time. While conventional linear systems theory would generally predict reliability and consistency at least to some degree, no consistent agreement within any particular set of value statements in CAS would be expected. Hence:

Hypothesis 2a: Each set of four items from the six attitude constructs are both internally consistent and consistent over time.

Hypothesis 2b: Each set of four items from the six attitude constructs are neither internally consistent and nor consistent over time.

3.1.3 Individual survey items

We also wanted to examine possible changes in individual items of the attitude survey on a longitudinal basis. We reasoned that if a survey item was related to the HPWS intervention, and if the item gained strength in responses between t2 and t3, then that item would likely be related to company responsibility. The question was, what would a conventional perspective predict about whether and which items would change, and how would these predictions vary in a Complexity approach? First, since all the items are associated with constructs that reflect the aims of CR, conventional theory would predict that all items would improve between t2 and t3 in a linear trend. Secondly, although study of attitudinal changes in complex systems is largely lacking in CR, we believe that survey items indicating individuals' active engagement and participation in the system would be expected to change more over time than passive expressions of ideas and beliefs. In sum:

Hypothesis 3a: All 24 items of the attitude survey will rise between t2 and t3.

Hypothesis 3b: Survey items that involve action and interaction will rise between t2 and t3.

4 RESULTS

4.1 Hypothesis 1

Hypothesis 1 was tested using the chi-square test of association. A non-parametric approach was called for due to the small and unequal sample sizes and the ordinal nature of the data. The chi square reveals the tendency of the modes of each assessment, which we considered a reasonable way to test the hypotheses.

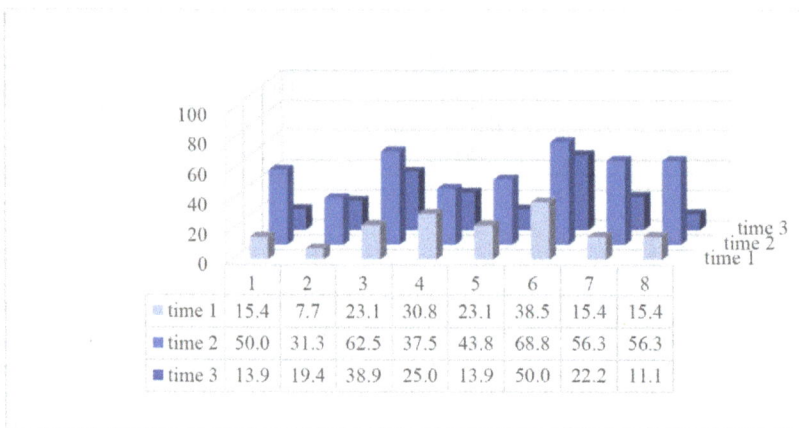

	1	2	3	4	5	6	7	8
time 1	15.4	7.7	23.1	30.8	23.1	38.5	15.4	15.4
time 2	50.0	31.3	62.5	37.5	43.8	68.8	56.3	56.3
time 3	13.9	19.4	38.9	25.0	13.9	50.0	22.2	11.1

Figure 1: Bar chart showing strength of positive responses in the Zinger survey at t1, t2, and t3.

Table 3: Zinger survey: Chi-square test of association with *p*-values and Cramer's V effect size.

Zinger Item	X^2	*df*	N	*p*	V
1 Creativity	19.994	6	65	0.003	0.392
2 Productivity	8.180	4	65	0.085	0.251
3 Collegiality	7.312	6	65	0.293	0.237
4 Collaboration	6.137	6	65	0.408	0.217
5 Job Satisfaction	10.386	4	65	0.034	0.283
6 Autonomy	7.674	6	65	0.263	0.243
7 Engagement	10.790	6	65	0.095	0.288
8 Positive Morale	19.401	6	65	0.004	0.386

Figure 1 shows an uptick in all Zinger feelings between t1 and t2, signaling a rise in these aspects of employee wellbeing as the HPWS project rolled out. This result agrees with the conventional approach and most of the literature that has found that employee wellbeing improves in the short run. Quite unexpected, however, were the results for t3, namely that all eight Zinger items then decreased at t3 relative to t2. Chi-square results (Table 3) show that this peaked pattern was significantly different than that which would be expected by chance for five of the eight Zinger feelings (shaded): creativity, productivity, job satisfaction, engagement, and morale. Using Cramer's V as a measure of effect size, the significant results showed a moderate effect which was only a bit larger than the effect seen with the remaining items. It suggests the possibility that there may be other significant effects too small to detect with the low power of our study.

4.2 Hypothesis 2

We found no significant measures of reliability for any of the assessed attitude constructs in our data. This result shows a complete lack of support for Hypothesis 2a and full support for Hypothesis 2b. To consider the alternative possibility that in fact there were attitudinal factors present, just not those we had predicted, we also conducted an exploratory factor analyses at both t2 and t3. There was indeed a factor structure in evidence at t2, but we could find little logic to connect the items involved. There was another but completely different factor structure at t3. Hence, we surmised that there was unlikely to be any meaningful overt or latent constructs in our data. Perhaps these data were confounded by the limitations of our sample, but these results may also inform revisions to our theories.

4.3 Hypothesis 3

Hypothesis 3 was analysed with the chi-square test, again due to the small and variable sample, and the use of an ordinal scale. Results shown in Table 3 indicate that 3 of the 24 items rose significantly between t2 and t3. These were:

#16: I don't buy consumer products if I know that unethical, unjust, or harmful practices were involved in their production;

#20: If I make plans, I generally succeed in executing them;

#21. I do get enough resources to be effective in my job (after reverse scoring).

Table 4: Attitude Survey: Chi-square test of association with p-values and Cramer's V effect
size.

Attitude Item	X^2	df	N	p	V
1	2.195	2	60	0.334	0.191
2	0.526	2	63	0.769	0.091
3	1.538	3	61	0.673	0.159
4	2.440	2	62	0.295	0.198
5	0.549	1	63	0.459	0.093
6	0.929	2	63	0.628	0.121
7	2.982	3	63	0.394	0.218
8	3.846	2	63	0.146	0.247
9	0.765	2	63	0.682	0.110
10	1.413	2	63	0.493	0.150
11	1.452	2	63	0.484	0.152
12	2.249	2	63	0.325	0.189
13	0.870	2	62	0.647	0.118
14	2.590	3	63	0.459	0.203
15	1.334	2	62	0.513	0.147
16	6.046	2	62	0.049	0.312
17	2.058	3	63	0.560	0.181
18	3.888	3	62	0.274	0.250
19	1.438	3	63	0.697	0.151
20	8.007	3	63	0.046	0.357
21	12.806	2	63	0.002	0.451
22	0.774	2	63	0.679	0.111
23	3.954	2	63	0.138	0.251
24	3.115	3	63	0.374	0.222

Hypothesis 3a was partially supported: 3 of the 24 individual survey items rose significantly from t2 to t3. It appears that there was not a strong enough effect of the HPWS project over time to influence all attitude items to change as predicted over the longer term of one year. Hypothesis 3b was rather more strongly supported but, in our view, due to a radically different logic. Hypothesis 3b was based on the argument that items indicating action and interaction would be most likely to exhibit a significant increase at t3. Of all the 24 items in the survey, we believe items 3, 7, 16, 20, 21, 22, 23, and 24 fall into this category. Therefore, the fact that items 16, 20, and 21 did increase significantly does lend support to the idea that action-oriented actors do have more impact in a complexity environment than those who may have equally strong beliefs without engaging others.

5 DISCUSSION AND CONCLUSION

Our study and data analysis have shown that conventional linear systems analysis was inadequate alone to explain internal sustainability adoption processes in the public sector organization we studied. Specifically, the spike-and-return pattern found in the longitudinal Zinger Survey results of Hypothesis 1 defies the expectation of ongoing linear improvement in the employee wellbeing over the long term. Also, the universality of the decline in wellbeing indicators at t3 begs further investigation. In Hypothesis 2, the complete absence of either expected or latent attitudes could be due to a failure of conventional approaches to yet identify instrumental feelings and attitudes in employee wellbeing, or it could be that the nonlinear social construction process is not well enough understood in complexity approaches to CSP to make accurate predictions. Supporting the latter possibility are Hypothesis 3 results, where the complexity perspective indicated the importance of social action and interaction as the source of productive systemic change, over and above tacit beliefs and attitudes. Linear systems theory had no alternative explanation for these results.

In sum, perhaps the most reasonable statement that can be made is that the taken-for-granted dominance of conventional systems perspectives in CR research needs to be rethought. The neoclassical, positivist, and reductionist model of employee behavior is certainly not the only and may not be the best way to study internal organizational shifts towards sustainability. Indeed, its exclusive use may actually obscure what we're now looking for, a more nuanced understanding of employees' idiosyncratic processes related to sustainability adoption. Complexity theory, new and largely untried in the CSR field to date, offers a compelling addition.

We propose that conventional and complexity approaches are distinct, incommensurable modes of company responsibility. CSR is derived in the conventional approach and refers to operations- and firm-level CR objectives such as resource optimization, initiative implementation, and achievement of all manner of tangible responsibility objectives. On the other hand, sustainability is derived in complexity theory and CAS. Sustainability is a guiding concept for the mutual health and resilience of the most macro-level systems, including the biosphere and human society and enterprise [23]. We associate this concept and ontology with 'sustainability thinking,' [24] which is the visioning, collaborating, and adapting that underlie the achievement of temporary markers of CSR progress. Pogutz and Winn [25] have coined the term, 'sustainability fit,' as the condition in which a company can 'adapt and align dynamically [i.e. coevolve] with the resilience of the ecosystem within which it is embedded.' We extend their notion to introduce the idea of 'sustainability fitness' via complex systems principles, to describe the mindset and capabilities for balanced adaptation in the ongoing search for earthwise sustainability. CSR can be managed and stewarded, but sustainability must be nurtured, in an ongoing co-operative process among equals.

5.1 Limitations and alternative explanations

There are several issues with our study. Although this longitudinal study achieved high response rates, the sample size was small and as such analysis was restricted to non-parametric methods. Despite the small power present, some significant results were found warranting further research with larger samples. The Hawthorne effect and regression to the mean may have been in play in the spike-and-return phenomenon. Also, what are called 'confounding factors' in positivist research, and considered 'as-yet unknown complex processes' in CAS, are likely to have had an effect on our results.

5.2 Future research

The project for future research is not only to tease out finer and finer antecedents, moderators, mediators, etc., in CSR models. This agenda is firmly and myopically situated in conventional positivist and reductionist ontology. It is equally important to operationalize complexity theory and apply it through a variety of methods to dig deeper into the 'black box' of human, idiosyncratic micro-processes in sustainability fitness. Finally, it is important to take the birds-eye view at times, seeking not to integrate, and therefore, obfuscate the unique perspectives offered by each theory, but to consider them simultaneously for informative insights and ideas [26]. In this way, the business and management academy can maintain and increase its relevance in this desperately crucial field.

REFERENCES

[1] Aguinis, H. & Glavas, A., What we know and don't know about corporate social responsibility: a review and research Agenda. *Journal of Management* **38**(4), pp. 932–968, 2012.
https://doi.org/10.1177/0149206311436079

[2] Applebaum, E., Bailey, T., Berg, P. & Kalleberg, A.L., *Manufacturing Advantage: Why High-Performance Work Systems Pay Off*. Ithica, NY: Cornell University Press, 2000.

[3] Jin Suh, Y., The role of relational social capital and communication in the relationship between csr and employee attitudes: a multilevel analysis. *Journal of Leadership & Organizational Studies,* **23**(4), pp. 1–14, 2016.

[4] Hawkin, P., Lovins, A. & Lovins, H., *Natural Capitalism: The Next Industrial Revolution,* London: Earthscan, 2010.

[5] Peloza, J., The challenge of measuring financial impacts from investments in corporate social performance. *Journal of Management,* **35**(6), pp. 1518–1541, 2009.
https://doi.org/10.1177/0149206309335188

[6] Wang, H., Corporate social responsibility: an overview and new research directions. *Academy of Management Journal,* **59**(2), pp. 534–544, 2016.
https://doi.org/10.5465/amj.2016.5001

[7] Margolis, J. & Walsh, J., Misery loves companies: rethinking social initiatives by business. *Administrative Science Quarterly,* **48**, pp. 268–305, 2003.
https://doi.org/10.2307/3556659

[8] Starik, M. & Kanashiro, P., Toward a theory of sustainability management: uncovering and integrating the nearly obvious. *Organization and Environment,* **26**(1), pp. 7–30, 2013.
https://doi.org/10.1177/1086026612474958

[9] Anderson, P., Complexity theory and organization science. *Organization Science,* **10**(3), pp. 216–232, 1999.
https://doi.org/10.1287/orsc.10.3.216

[10] McKelvey, B., Quasi-natural organization science. *Organization Science,* **8**(4), pp. 352–380, 1997.
https://doi.org/10.1287/orsc.8.4.351

[11] Porter, T., Coevolution as a research framework for organizations and the natural environment. *Organization and Environment,* **19**(4), pp. 479–504, 2006.
https://doi.org/10.1177/1086026606294958

[12] Robone, A., Jones, A. & Rice, N., Contractural conditions, working conditions and their impact on health and well-being. *European Journal of Health Economics,* **12**, pp. 429–444, 2011.
https://doi.org/10.1007/s10198-010-0256-0

[13] Pfeffer, J., *The Human Equation: Building Profits by Putting People First*, Cambridge, MA: Harvard Business Press, 1998.

[14] Huselid, M., The impact of human resource management practices on turnover, productivity, and corporate financial performance. *Academy of Management Journal*, **38**(3), pp. 635–672, 1995.
https://doi.org/10.2307/256741

[15] Rothenberg, S., Hull, C. & Tang, C., The impact of human resource management on corporate social performance strengths and concerns. *Business and Society*, pp. 1–28, 2015.

[16] Oana, G. & Shahrazad, H'., Does civil society create social entrepreneurs?. In *Ideas, ed. Federal Reserve Bank of St. Louis*, 2013, available from http://anale.steconomiceuoradea.ro/volume/2013/n1/069.pdf.

[17] Sheehan, M., Garavan, T. & Carbery, R., Innovation and human resource development (HRD). *European Journal of Training and Development*, **38**(1/2), pp. 2–14, 2014.
https://doi.org/10.1108/EJTD-11-2013-0128

[18] Barnett, M., Stakeholder influence capacity and the variability of financial returns to corporate social responsibility. *Academy of Management Review* **32**(3), pp. 794–816, 2007.
https://doi.org/10.5465/AMR.2007.25275520

[19] Combs, J., Hall, A. & Ketchen, D., How much do high-performance work practices matter? a meta-analysis of their effects on organizational performance. *Personnel Psychology*, **59**(3), pp. 501–528, 2006.
https://doi.org/10.1111/j.1744-6570.2006.00045.x

[20] Ko, J. & Smith-Walter, A., The relationship between hrm practices and organizational performance in the public sector: focusing on mediating roles of work attitudes. *International Review of Public Administration*, **18**(3), pp. 209–231, 2013.
https://doi.org/10.1080/12294659.2013.10805270

[21] Hull, C. & Rothenberg, S., Firm performance: the interactions of corporate social performance with innovation and industry differentiation. *Strategic Management Journal*, **29**, pp. 781–789.
https://doi.org/10.1002/smj.675

[22] Stacey, R., *Complex Responsive Processes in Organizations: Learning and Knowledge Creation*, London: Routledge, 2001.

[23] Waddock, S., *Leading Corporate Citizens: Vision, Values, Value-Added*, 3 edn., New York: McGraw Hill, 2009.

[24] Porter, T. & Derry, R., Sustainability and business in a complex world. *Business and Society Review*, **117**(1), pp. 33–54, 2012.
https://doi.org/10.1111/j.1467-8594.2012.00398.x

[25] Pogutz, S. & Winn, M., Cultivating ecological knowledge for corporate sustainability: barilla's innovative approach to sustainable farming. *Business Strategy and the Environment*, **25**(6), pp. 435–448, 2016.
https://doi.org/10.1002/bse.1916

[26] Espinoza, A. & Porter, T., Sustainability, complexity and learning: insights from complex systems approaches. *The Learning Organization*, **18**(1), pp. 54–72, 2011.
https://doi.org/10.1108/09696471111096000

TWO COMPLEX ADAPTIVE SYSTEMS IN HUMAN DEVELOPMENT: FURTHER STUDIES OF DENTAL AND FINGERPRINT PARAMETERS

R.J.O. TADURAN[1], S. RANJITKAR[1], T. HUGHES[1], G. TOWNSEND[1] & A.H. BROOK[1,2]
[1]Craniofacial Biology Research Group, Adelaide Dental School, University of Adelaide, Australia.
[2]School of Dentistry, Queen Mary University, London, UK.

ABSTRACT
This paper reports further results and an extension of the study presented at Complex Systems 2016. Human teeth and fingerprints both arise from genetic/epigenetic/environmental interactions and have embryological pathways with epithelial–mesenchymal interactions. The aims of this study were to determine the nature and extent of sexual dimorphism in teeth and fingerprints of twins at two different ages and to explore whether both systems display the features of complex adaptive systems. Buccolingual (BL) measurements from both primary and permanent teeth and ridge breadth (RB) measurements from fingerprints of the same set of Australian twins (28 males and 31 females aged 8 to 10 years, and aged 13 to 16 years, respectively) were collected and analysed. Sexual dimorphism was observed in both the primary and permanent dentitions, with the latter showing greater differences than the former. There was no observed sexual dimorphism in the fingerprints at 8 to 10 years. However, a few fingers (left index, left ring, and right middle) at 13 to 16-years exhibited significant differences, suggesting that friction ridges expand over time. It is concluded that both the dentition and dermatoglyphics display sexual dimorphism and characteristics of complex adaptive systems.
Keywords: buccolingual, complex adaptive system, dentition, dermatoglyphics, fingerprints, human development, ridge breadth, sexual dimorphism, tooth size

1 INTRODUCTION
Sexual dimorphism is the difference between sexes of the same biological species in phenotype or appearance. Some researchers have proposed that sexual differences are regulated by sex chromosomes [1, 2], but there are some who have suggested that hormonal influences are also important [3, 4]. Sexual dimorphism in human dentition and dermatoglyphs have been studied separately, and results are fairly consistent: males have larger tooth crown diameters than females [5, 6], sexual dimorphism is greater in permanent than in primary teeth [6, 7] and adult males have fewer finger ridges than females [8, 9].

Human development is a complex adaptive process that is influenced by genetic, epigenetic and environmental factors [10]. Genes interact with epigenetic and environmental elements and create complex networks within cells, and from this process the higher level tissues are formed. During embryonic growth, patterning, or the establishment of groups of cells in the proper relationship to each other and to surrounding tissues, occurs. Patterning is a longitudinal event that eventually leads to differentiation of cells to assume specialised functions and shapes.

The development of the human dentition and dermatoglyphs has similar embryological origin from epithelial–mesenchymal interactions [11]. Primary teeth commence development around 4 to 6 weeks in utero [10], while ridged skin on the fingers starts to form around 10 to 16 weeks in utero [12]. Once the patterns have been stabilised, their unique and persistent morphology makes them valuable models in studying sexual dimorphism. Only one study has explored possible correlations between the human dentition and dermatoglyphs in sub-adult Australian twins, and sexual dimorphism was observed in both primary and permanent teeth but not in fingerprints [13]. This study aimed to determine the nature and extent of sexual dimorphism in teeth and fingerprints of twins and explore whether both systems display the features of complex adaptive systems.

2 MATERIALS AND METHODS

Twin samples were acquired from the ongoing longitudinal research of the Craniofacial Biology Research Group in the Adelaide Dental School at the University of Adelaide [14], which is one of the four most extensive studies of its type in the world [15]. Serial casts of primary and permanent teeth, and rolled ink fingerprints of individuals aged 8 to 10 years and 13 to 16 years from a single cohort of monozygotic and dizygotic Australian twins (28 males and 31 females) were gathered and analysed. Dental casts showing wear, caries, or restorations and ten-prints with smudged ink and scarred patterns in any of the fingerprints were excluded.

Buccolingual crown diameter (BL) was measured as the breadth or distance between the buccal/labial and lingual surfaces of the crown [16, 17] by using a 2D imaging system. Using an adjustable stage, dental casts were orientated to obtain the correct plane or angle before obtaining images and calibrated Image J [18] software was used to digitise landmarks (Fig. 1). Measurements were obtained for central incisors (I1), lateral incisors (I2), canines (C), first molars (M1) and second molars (M2) of primary and permanent teeth.

Ridge breadth (RB) was determined by measuring the distance of 10 parallel ridges with no obstruction such as scars or white creases and/or interfering minutiae such as bifurcations, ridge endings, and short ridges. Measurements began and ended with valleys, or the spaces before the first ridge and after the tenth ridge (Fig. 2). This method is independent and not influenced by fingerprint pattern type and finger area [19].

Data were statistically analysed using XLSTAT statistical software. Descriptive statistics including means, standard deviations (SD) and coefficients of variation (CV) were computed

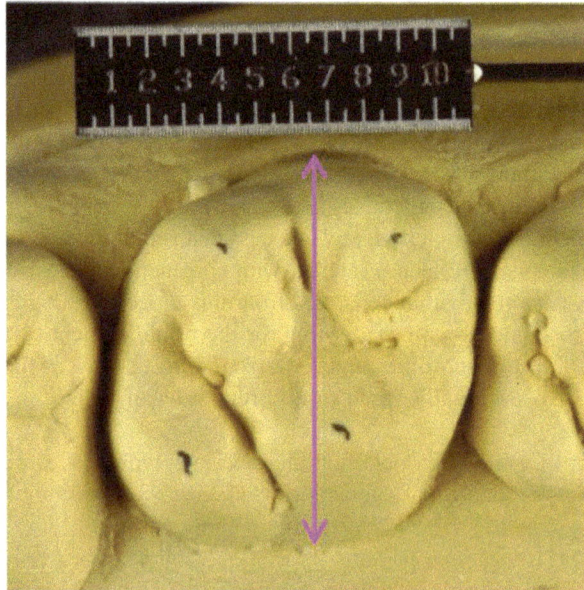

Figure 1: Buccolingual (BL) measurement of a permanent upper first molar from the occlusal view.

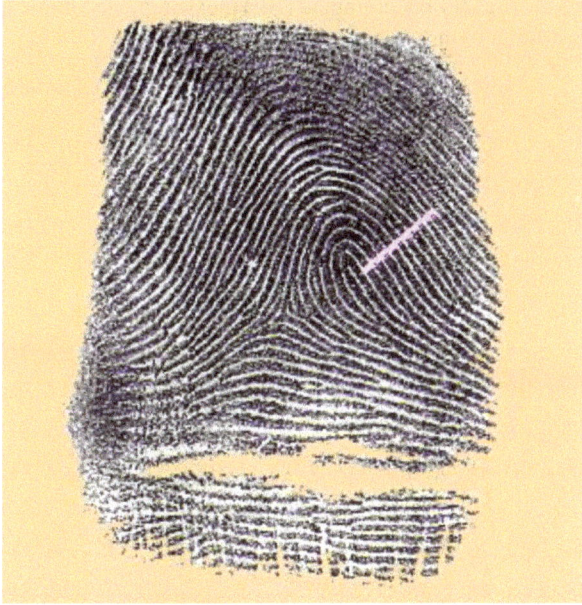

Figure 2: Sample of Ridge Breadth (RB) measurement.

for BL and RB variables. Differences between sexes and sides were calculated using Student's unpaired t-test. RB differences between age groups were compared with paired t-tests, and differences among fingers were examined with analysis of variance (ANOVA). Pearson's correlation coefficient was calculated to examine the strength of associations between the variables.

3 RESULTS

BL and RB measurements were found to be normally distributed, and results of intra- and inter-operator repeatability tests determined that errors in measurements were negligible and not likely to bias the results. Shown in Table 1 are the mean values, SD and CV of buccolingual (BL) measurements of primary and permanent teeth.

Highlighted in yellow are the sexually dimorphic dental measurements, where mean values are different between sexes at $p < 0.05$. Mean values of BL crown sizes of males were consistently greater compared to females in all types of teeth. Permanent dentitions showed greater sexual dimorphism compared to primary dentitions. There were no left-right differences observed in BL measurements of all primary and permanent teeth.

Shown in Table 2 are the mean values, SD and CV of ridge breadth (RB) of fingerprints of 8 to 10 year-old cohort and 13 to 16 year-old cohort.

All mean values of RB were statistically different to each other at $p < 0.05$. Highlighted in blue are the RB means that were found to be statistically different on both sides. More left-right differences were observed in the younger (8 to 10 years old) cohort. Most fingers were asymmetric in both sexes, except for the index fingers and thumbs in males and index fingers in females. It was observed that fingers on the right side consistently have greater RB, which indicates thicker finger ridges.

Table 1: Descriptive statistics for buccolingual (BL) measurements of primary and permanent teeth of Australian twins.

		Males								Females						
		Right				Left				Right				Left		
	n	Mean	SD	CV (%)	n	Mean	SD	CV (%)	n	Mean	SD	CV (%)	n	Mean	SD	CV (%)
							Primary									
							Maxillary									
i1	28	5.07	0.33	6.44	28	5.12	0.33	6.38	31	4.92	0.34	6.97	31	4.96	0.36	7.25
c	28	6.20	0.42	6.81	28	6.18	0.42	6.87	31	6.10	0.39	6.41	31	6.08	0.38	6.23
m1	28	8.79	0.43	4.88	28	8.76	0.42	4.77	31	8.55	0.34	3.96	31	8.54	0.34	4.01
m2	28	10.00	0.48	4.76	28	9.96	0.43	4.37	31	9.68	0.40	4.12	31	9.64	0.39	4.02
							Mandibular									
i1	28	3.88	0.30	7.62	28	3.83	0.25	6.40	31	3.72	0.27	7.30	31	3.71	0.24	6.58
i2	28	4.40	0.33	7.48	28	4.38	0.31	6.99	31	4.28	0.29	6.71	31	4.29	0.28	6.60
c	28	5.65	0.36	6.40	28	5.65	0.36	6.32	31	5.58	0.38	6.90	31	5.58	0.35	6.29
m1	28	7.09	0.38	5.31	28	7.17	0.37	5.16	31	6.86	0.40	5.77	31	6.94	0.37	5.29
m2	28	8.72	0.38	4.36	28	8.72	0.38	4.35	31	8.38	0.40	4.82	31	8.41	0.37	4.36
							Permanent									
							Maxillary									
I1	28	7.27	0.56	7.69	28	7.29	0.55	7.55	31	7.04	0.55	7.75	31	7.04	0.56	7.93
C	28	8.32	0.56	6.73	28	8.41	0.61	7.28	31	7.91	0.54	6.81	31	7.97	0.56	6.96
M1	28	11.79	0.56	4.79	28	11.73	0.54	4.60	31	11.22	0.53	4.75	31	11.16	0.50	4.48
M2	28	11.95	0.70	5.88	28	12.07	0.84	7.00	31	11.18	0.69	6.14	31	11.07	0.61	5.48
							Mandibular									
I1	28	6.19	0.46	7.41	28	6.13	0.51	8.34	31	5.93	0.46	7.74	31	5.97	0.43	7.23
I2	28	6.47	0.53	8.23	28	6.41	0.55	8.51	31	6.25	0.53	8.41	31	6.28	0.47	7.47
C	28	7.66	0.65	8.50	28	7.66	0.65	8.45	31	7.20	0.48	6.64	31	7.29	0.57	7.77
M1	28	10.54	0.47	4.48	28	10.56	0.50	4.70	31	9.99	0.48	4.78	31	10.07	0.48	4.77
M2	28	10.70	0.58	5.39	28	10.64	0.61	5.70	31	10.01	0.61	6.12	31	10.07	0.57	5.68

Highlighted in yellow are the sexually dimorphic RB measurements, where mean values are different between sexes at $p < 0.05$. Left index, right middle and left ring fingers were observed to exhibit male-female differences in the older cohort (13 to 16 years old), with greater mean values for RB in males, which indicates thicker friction ridges. Based on paired t-test, all mean values of RB are different between age groups at $p < 0.05$, with the older cohort having greater RB values compared to the younger group.

Table 2: Descriptive statistics for ridge breadth (RB) of fingerprints of Australian twins.

| | Males | | | | | | | | Females | | | | | | | |
| | Right | | | | Left | | | | Right | | | | Left | | | |
	n	Mean	SD	CV (%)	n	Mean	SD	CV (%)	n	Mean	SD	CV (%)	n	Mean	SD	CV (%)
8–10 years old																
Thumb	28	4.24	0.54	12.66	28	4.21	0.57	13.59	31	4.32	0.58	13.39	31	4.07	0.46	11.24
Index	28	3.98	0.45	11.21	28	3.97	0.60	15.19	31	3.98	0.46	11.58	31	3.94	0.48	12.22
Middle	28	3.92	0.51	13.07	28	3.76	0.49	13.15	31	3.93	0.50	12.77	31	3.69	0.50	13.65
Ring	28	3.83	0.54	14.09	28	3.49	0.45	12.98	31	3.83	0.44	11.46	31	3.46	0.45	13.16
Little	28	3.80	0.54	14.23	28	3.61	0.48	13.37	31	3.82	0.47	12.24	31	3.67	0.42	11.39
13–16 years old																
Thumb	28	4.69	0.54	11.58	28	4.68	0.55	11.79	31	4.61	0.59	12.90	31	4.55	0.52	11.39
Index	28	4.40	0.55	12.60	28	4.62	0.56	12.19	31	4.36	0.65	15.00	31	4.30	0.53	12.29
Middle	28	4.29	0.60	13.99	28	4.05	0.41	10.07	31	4.03	0.39	9.72	31	3.89	0.50	12.77
Ring	28	4.12	0.52	12.57	28	3.88	0.45	11.72	31	3.94	0.51	12.87	31	3.64	0.43	11.72
Little	28	4.16	0.53	12.64	28	4.06	0.51	12.48	31	4.09	0.38	9.28	31	3.95	0.40	10.22

Pearson's coefficients (r) between teeth and fingerprints are presented in Table 3. Highlighted in yellow are the statistically significant correlations between dental characteristic, BL, and dermatoglyphic trait, RB at $p < 0.05$. In general, the correlations between teeth and fingerprints are statistically significant but low in magnitude.

Correlation coefficients were calculated within groups of dental and dermatoglyphic variables with significance set at $p < 0.05$. All BL measurements taken from different tooth types were positively correlated to each other in the primary teeth (0.32 to 0.92). Meanwhile, only some BL diameters (128 of 306 in males, 226 of 306 in females) were positively correlated to each other in the permanent teeth (0.36 to 0.94), and more significant values were observed in females than males. On the other hand, only some RB measurements from different fingers were positively correlated to each other (90 of 90 in young males and 83 of 90 in young females, 64 of 90 in old males and 40 of 90 in old females), with more significant values in males, and all values are significant in the young cohort males. Greater r values were observed in RB at an older age (0.39 to 0.76) compared to the younger age (0.21 to 0.67).

Table 3: Pearson correlation coefficients of BL (buccolingual width) and RB (ridge breadth).

| | Primary | | Permanent | |
Ridge Breadth	Maxillary	Mandibular	Maxillary	Mandibular
Males	0.28	0.30	0.25	0.35
Females	0.28	0.29	0.24	0.37

4 DISCUSSION

The degree and patterning of sexual dimorphism in the dentition varies according to tooth type. Our observation of the permanent dentition showing more pronounced sexual dimorphism than the primary dentition agrees with previous findings [4, 13]. The permanent molars displayed the largest sexual dimorphism in BL measurements, similar to previous studies [20, 21]. It has been suggested that dental development might occur under relatively high levels of testosterone influence [4], and this could explain the differences in sexual dimorphism between primary and permanent teeth of same individuals.

The degree and patterning of sexual dimorphism in the dermatoglyphs varies according to finger type and side. In this study, there was no observed sexual dimorphism at the age of 8 to 10 years, while fingerprints at 13 to 16 years of age displayed sexual dimorphism in the left index and ring fingers, and right middle finger. Our results are consistent with our previous study and support the idea that friction ridges expand as individuals grow and develop, probably more in males than females [13]. Sexual dimorphism in dermatoglyphic development seems to be initiated during puberty, when a testosterone surge occurs in males [22].

In normal male development, three surges of testosterone occur: the first surge happens at around the 7th to 9th week of pregnancy, and the testosterone level is highest around the 14th week following testicular differentiation [23, 24]; the second surge initiates after birth because of the reduction of oestrogen produced by the placenta [22]; and the third surge occurs during puberty. Meanwhile, primary teeth begin to form at around 4 to 6 weeks in utero [11] until around one year after birth. Permanent teeth commence development 14 weeks in utero and around 14 years of age [25]. On the other hand, primary ridges start to form at around 10 to 16 weeks and end on the 17th week, then secondary ridges develop until the 24th week in utero [12]. Our results are consistent with our previous study [13] and further support the idea that the first two testosterone surges have a critical role in the sexual dimorphism of both the primary and permanent teeth, while the third testosterone surge strongly influences the sexual dimorphism of fingerprints.

The human body is a complex adaptive system, and human development is a complex adaptive process [10]. This research has shown that both teeth and fingerprints are interconnected, yet they still have a degree of autonomy. They share a similar embryological origin and epithelial–mesenchymal interactions [11], yet they develop and interact with epigenetic and environmental factors differently. The interactions may be unpredictable, with no central control, but they are not random, as regularities and patterns emerge to find the best fit with the environment.

This research furthers the investigation on the complex mechanisms and interactions occurring during dental, dermatoglyphic and general development with buccolingual (BL) measurements of the teeth and ridge breadth (RB) measurements of the fingerprints. It is our second attempt to study both human dental and dermatoglyphic traits. Most studies have been conducted on the human dentition and dermatoglyphs separately, and no effort has been made to explore possible correlations between the two.

ACKNOWLEDGEMENTS

The authors wish to thank The Australian Dental Research Foundation (ADRF), the NHMRC of Australia, Australian Twin Registry and Australian Multiple Birth Association.

REFERENCES

[1] Guatelli-Steinberg, D., Sciulli, P.W. & Betsinger, T.K., Dental crown size and sex hormone concentrations: another look at the development of sexual dimorphism. *American Journal of Physical Anthropology*, **137**, pp. 324–333, 2008.

[2] Alvesalo, L., Human sex chromosomes in oral and craniofacial growth. *Archives of Oral Biology*, **54S**, pp. 18–24, 2009.

[3] Dempsey, P.J., Townsend, G.C. & Richards, L.C. Increased tooth crown size in females with twin brothers: evidence for hormonal diffusion between human twins in utero. *American Journal of Human Genetics*, **11**, pp. 577–586, 1999.

[4] Ribeiro, D.C., Brook, A.H., Hughes, T.E., Sampson, W.J. & Townsend, G.C., Intrauterine hormone effects on tooth dimensions. *Journal of Dental Research*, **92**, pp. 425–31, 2013.

[5] Moorrees, C.F.A., Thomsen, S.O., Jensen, E. & Yen, P.K., Mesiodistal crown diameters of the deciduous and permanent teeth in individuals. *Journal of Dental Research*, **36**, pp. 39–47, 1957.

[6] Ribeiro, D., Sampson, W., Hughes, T., Brook, A. & Townsend, G., *Sexual dimorphism in the primary and permanent dentitions of twins: an approach to clarifying the role of hormonal factors* (Chapter 5). New Directions in Dental Anthropology: Paradigms, Methodologies and Outcomes, ed. G. Townsend, E. Kanazawa, H. Takayama, University of Adelaide Press: South Australia, pp. 46–64, 2012.

[7] Garn, S.M., Lewis, A.B., Swindler, D.R. & Kerewsky, R.S., Genetic control of sexual dimorphism in tooth size. *Journal of Dental Research*, **46**, pp. 963–972, 1967.

[8] Acree, M.A., Is there a gender difference in fingerprint ridge density? *Forensic Science International*, **102**, pp. 35–44, 1999.

[9] Taduran, R.J.O., Tadeo, A.K.V., Escalona, N.A.C., & Townsend, G.C., Sex determination from fingerprint ridge density and white line counts in Filipinos. *HOMO - Journal of Comparative Human Biology*, **67**, pp. 163–171, 2016.

[10] Brook, A.H., Brook O'Donnell, M., Hone, A., Hart, E., Hughes, T.E., Smith, R.N. & Townsend, G.C., General and craniofacial development are complex adaptive processes influenced by diversity. *Australian Dental Journal*, **59S**, pp. 13–22, 2014.

[11] Nanci, A., *Ten Cate's oral histology: development, structure, and function*, 7th edition, Elsevier Health Sciences: Missouri, pp. 16–107, 2008.

[12] Kücken, M. & Newell, A.C., Fingerprint formation. *Journal of Theoretical Biology*, **235**, pp. 71–83, 2005.

[13] Taduran, R.J.O., Ranjitkar, S., Hughes, T., Townsend, G. & Brook, A.H., Complex systems in human development: sexual dimorphism in teeth and fingerprints of Australian twins. *International Journal of Design & Nature and Ecodynamics*, **11**, pp. 676–685, 2016.

[14] Townsend, G., Bockmann, M., Hughes, T., Mihailidis, S., Seow, K.W. & Brook, A., *New approaches to dental anthropology based on the study of twins* (Chapter 2). New Directions in Dental Anthropology: Paradigms, Methodologies and Outcomes, ed. G. Townsend, E. Kanazawa, H. Takayama, University of Adelaide Press: South Australia, pp. 10–21, 2012.

[15] Hughes, T.E., Townsend, G.C., Pinkerton, S.K., Bockmann, M.R., Seow, W.K., Brook, A.H., Richards, L.C., Mihailidis, S., Ranjitkar, S. & Lekkas, D., The teeth and faces of twins: providing insights into dentofacial development and oral health for practicing oral health professionals. *Australian Dental Journal*, **59S**, pp. 101–116, 2014.

[16] Brook, A.H., Griffin, R.C., Townsend, G., Levisianos, Y., Russell, J. & Smith, R.N., Variability and patterning in permanent tooth size of four human ethnic groups. *Archives of Oral Biology*, **54S**, pp. S79–S85, 2009.

[17] Brook, A.H., Smith, R.N., Elcock, C., al-Sharood, M.H., Shah, A.A., Khalaf, F., Robinson, D.L., Lath, D.L. & Karmo, M., *The measurement of tooth morphology: validation of an image analysis system*. 13th International Symposium of Dental Morphology. E. Zadzinska, University of Lodz Press: Lodz, pp. 475–482, 2005.

[18] Schneider, C.A., Rasband, W.S. & Eliceiri, K.W., NIH Image to ImageJ: 25 years of image analysis. *Nature Methods*, **9**, pp. 671–675, 2012.

[19] Mundorff, A.Z., Bartelink, E.J. & Murad, T.A., Sexual dimorphism in finger ridge breadth measurements: a tool for sex estimation from fingerprints. *Journal of Forensic Sciences*, **59**, pp. 891–897, 2014.

[20] Ządzínska, E., Karasínska, M., Jedrychowska-Dánska, K., Watala, C. & Witas, H.W., Sex diagnosis of subadult specimens from Medieval Polish archaeological sites: metric analysis of deciduous dentition. *HOMO - Journal of Comparative Human Biology*, **59**, pp. 175–187, 2008.

[21] Girija, K. & Ambika, M., Permanent maxillary first molars: role in gender determination (morphometric analysis). *Journal of Forensic Dental Sciences*, **4**, pp. 101–102, 2012.

[22] Griffin, J.E. & Wilson, J.D., *Disorders of the testes and the male reproductive tract* (Chapter 18). Williams Textbook of Endocrinology, 10th edn., eds. P.R. Larsen, H.M. Kronemberg, S. Melmed & K.S. Polonsky, W.B. Saunders Company: Philadelphia, pp. 709–770, 2003.

[23] Reyes, F.I., Boroditsky, R.S., Winter, J.S.D. & Fairman, C., Studies on human sexual development. II. Fetal and maternal serum gonadotropin and sex steroid concentrations. *Journal of Clinical Endocrinology and Metabolism*, **38**, pp. 612–617, 1974.

[24] Knickmeyer, R.C. & Baron-Cohen, S., Fetal testosterone and sex differences. *Early Human Development*, **82**, pp. 755–760, 2006.

[25] AlQahtani, S.J., Hector, M.P. & Liversidge, H.M., Brief communication: the London atlas of human tooth development and eruption. *American Journal of Physical Anthropology*, **142**, pp. 481–490, 2010.

DELAY IN DENTAL DEVELOPMENT AND VARIATIONS IN ROOT MORPHOLOGY ARE OUTCOMES OF THE COMPLEX ADAPTIVE SYSTEM ASSOCIATED WITH THE NUMERICAL VARIATION OF HYPODONTIA

LEO CHEN[1], HELEN LIVERSIDGE[1], KE CHEN[2], MAURO FARELLA[3], SADAF SASSANI[4],
DILAN PATEL[4], AZZA AL-ANI[4] & ALAN BROOK[4]
[1]Queen Mary University, London.
[2]University of Liverpool, England.
[3]University of Otago, Otago, New Zealand.
4) University of Adelaide, South Australia.

ABSTRACT

The development of the dentition has the general characteristics of a complex adaptive system. Hypodontia is a developmental variation with not only a reduced number of teeth but also the teeth formed are smaller in size, have different crown and root morphology and are delayed in development. We have formed a multicentre, multidisciplinary collaborative study to investigate this complex system from its initiation to its outcome: from genotype with genetic/epigenetic/ environmental interactions to the accurate measurement of the phenotypic outcome. This paper reports an initial study of the root morphology and dental age component of the phenotype of the hypodontia patient compared to controls. The sample consists of orthodontic patients, 30 males and 30 females with hypodontia and 60 controls matched for age, gender and ethnicity. The material studied is the orthopantomographic radiograph of each patient. From these the number and site of each congenitally missing tooth is recorded. The number and shape of roots of each formed tooth are scored. Using the MATLAB computer programming platform, the distance between specific points on the crown and root, their area and hence the crown/ root ratio is computed, and the stages of dental development of each tooth scored; the degree of root development of the second permanent molar is particularly valuable in comparing between hypodontia patients and controls. By combining investigations from different stages of this biological complex adaptive system, we are using dental development, which is an accessible, non-invasive and accurately measurable paradigm to increase understanding of general development.
Keywords: Complex system, developments measures, hypodontia, radiographs.

1 INTRODUCTION

Developmental systems in biology may well behave as complex systems. It has been shown by Brook and O'Donnell [1] and Brook *et al.* [2] that dental development has the general characteristics of a complex adaptive system and that it is an accessible, non-invasive model for general development. The opportunity has arisen to examine the developmental process in relation to a specific variation of the number of teeth formed during development in some individuals. In hypodontia, together with the congenital absence of some teeth, those that are formed are smaller in size, have different crown and root morphology, and are delayed in development. We have formed a multicentre, multidisciplinary collaborative study to investigate this complex system from its initiation to its outcome: that is from the genotype with genetic/epigenetic/environmental interactions to the accurate measurement of the phenotypic outcome.

The aim of this part of the overall study is to investigate the root morphology and dental age component of the phenotype of the hypodontia patients compared to the controls.

© 2018 WIT Press, www.witpress.com
DOI: 10.2495/DNE-V13-N1-101-106

2 MATERIALS, PATIENTS AND METHODS

The sample consists of 120 orthodontic patients, 30 males and 30 females with hypodontia and 60 controls matched for age, gender and ethnicity. The age range of the patients was: hypodontia – males, 10.2 to 21.0 years; females, 10.8 to 19.3 years and controls – males, 10.1 to 17.3 years; females, 10.6 to 19.2 years. The material studied is the orthopantomographic radiograph of each patient. From these the number and site of each congenitally missing tooth in each patient is recorded.

The number and shape of the roots of each formed tooth is then scored. Using the MAT-LAB computer programming platform, we have developed programs to compute the distance between specific points on the crown and root, their area and hence the crown/root ratio is computed.

The stages of dental development of each tooth are also scored; the degree of root development of the second permanent molar is particularly valuable in comparing between these hypodontia patients and the controls.

3 METHOD DEVELOPMENT/DISCUSSION

The number of congenitally missing teeth is an indication of the severity of the condition and often varies between individuals affected by the same genetic mutation and even in the same family (Brook *et al.* [2]). This is an indication of the complex aetiology of the condition.

The site of the congenitally missing found in previous studies is not random, but relates to position in the dental arch and timing of development [3]. Those teeth that are positioned furthest distally and are the last to develop in their tooth type (incisor/premolar/molar) are those most frequently absent [4, 5]. This emphasises why, as an integral part of the overall study, in this component we are investigating timing of dental development, which in the complex developmental process may be related to congenital absence as well as smaller size and different crown and root shape.

Recording the distance between specific landmarks is undertaken using the computer programming platform MATLAB which can accept and display an X-ray image in any commonly used format. After highlighting two different landmarks, our program then computes the distance and displays on the screen as well as in the command window. It allows continuous pairwise measurements to avoid restarting the procedure. For anterior teeth, the distance between the Cement Enamel Junction and the root apex is measured [6]. This allows the crown-to-root ratio to be determined; this approach compensates for different tooth angulations between individuals.

For posterior teeth, the method developed by Holt and Brook [7] is used to determine the degree of Taurodontism. The axis between the mesial and distal aspects of the Cement Enamel Junction is taken as the baseline and the distances are measured between this line and the floor of the pulp chamber and the apex of the distal root (Fig. 1). There are conflicting findings in the literature concerning any association between Hypodontia and Taurodontism, possibly related to the failure in some studies to use clear, objective criteria to diagnose Taurodontism.

In addition, we are recording three rooted mandibular first molar teeth, fused molar roots, cuneiform molar roots and invaginations. These additional criteria were also monitored in a study regarding variations of tooth root morphology in a Romano-British population by Brook and Scheers [8].

The MATLAB program has also been adapted to calculate surface area of both the whole tooth as well as the crown and roots separately. These measurements are made by the operator selecting a starting landmark on the radiograph with the mouse and tracing the outline of the

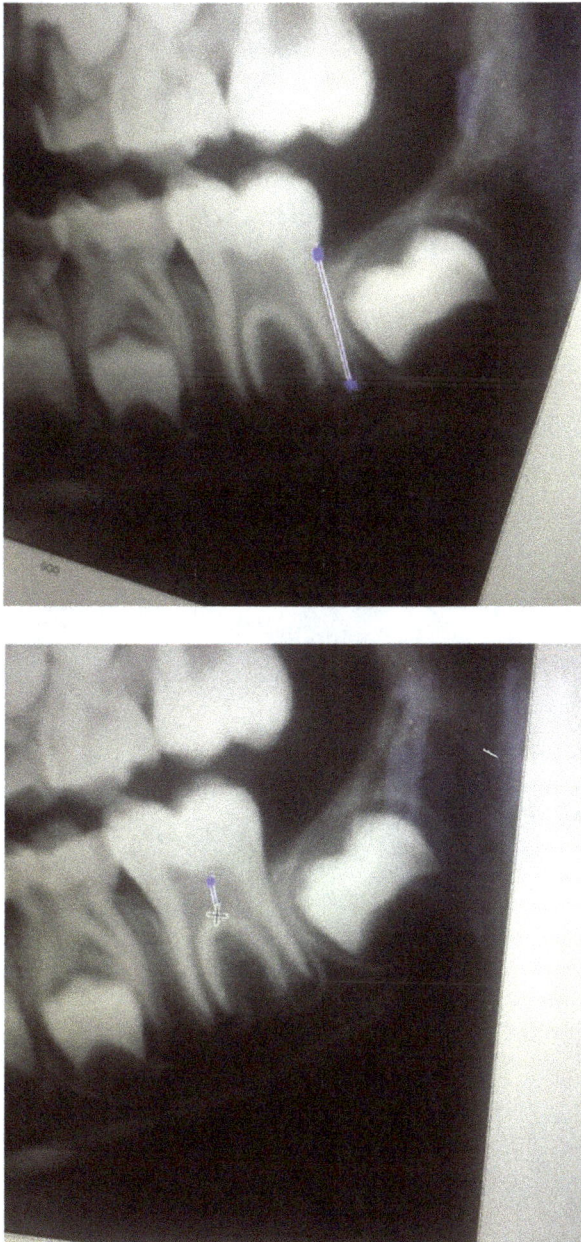

Figure 1: Top: Using the MATLAB system to measure the distance between the Cement Enamel Junction and the apex of the distal root. Bottom: The distance between the Cement Enamel Junction and the floor of the pulp chamber.

area to be measured before clicking again to stop at the end point. The crosshair used to outline the area is seen below in Figure 2 with the extracted area shown in Figure 3.

The program is designed to facilitate the measurement in an automatic way. Since there is possible error margin due to operator using this method, we have built in a curve smoothing

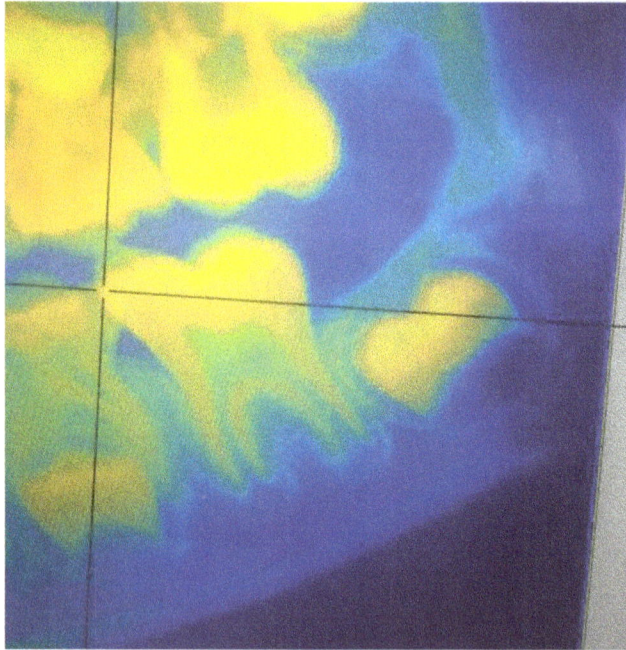

Figure 2: Start of the operator input.

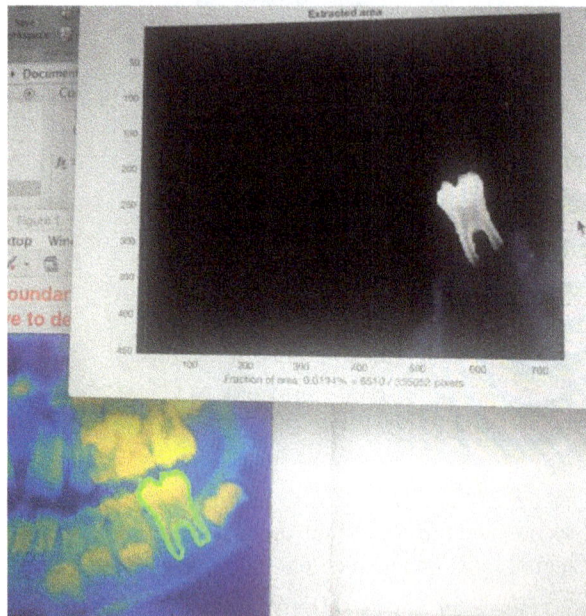

Figure 3: The traced area is then extracted and measured separately by the program.

function to automatically correct the input so that the final area calculated will be accurate as far as removal of noise or minor hand shaking is concerned. While this correction cannot compensate for major input error, the accuracy of input will be investigated by repeat measurements for intra-operator and inter-operator reproducibility.

This correction procedure can be statistically assessed to inform its reliability using boxplot (Fig. 4). The results from this section allow for the crown–to-root ratio to be measured by a novel approach, which can be compared against compared against existing techniques for accuracy and efficiently.

The program is further being used to measure root angle to gather more information and increase understanding of root morphology. This was perhaps the most challenging part of the study as there was some debate as to the angle we should be measuring. One consideration was

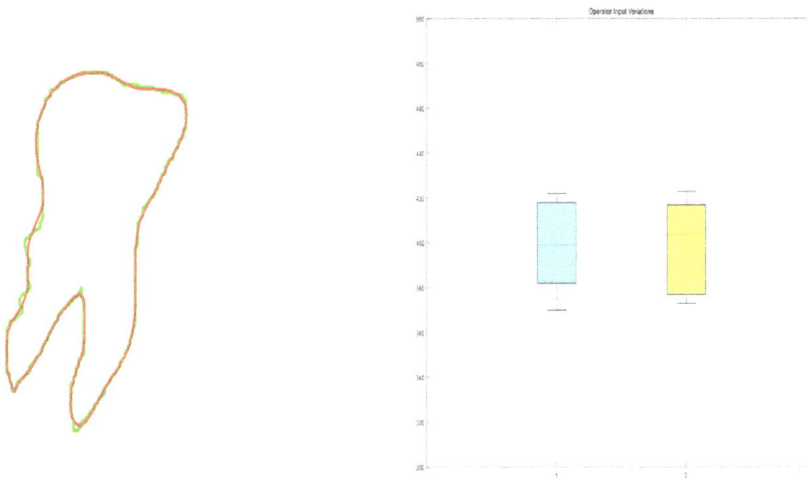

Figure 4: **Left**: Automatic curve smoothing (red) to reduce input errors (green).**Right**: Test of input variations (left cyan box for input and right yellow box for corrected values).

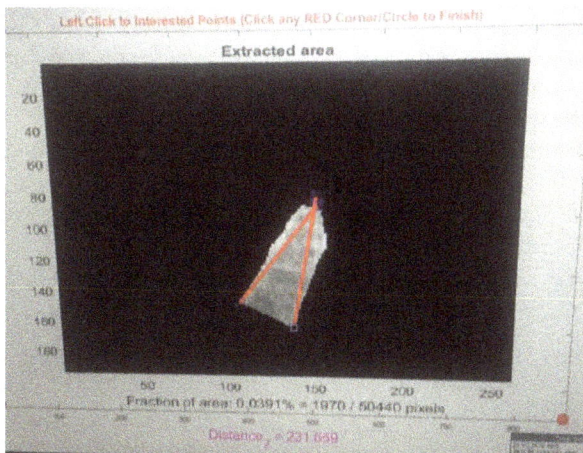

Figure 5: A simple example of operator defined angle. Other automatic ways are also considered.

to measure the root cone angle which was mentioned by Falk [9]. The method we have developed included selecting the most apical portion of the root and using the angle created by this point and the most coronal mesial and distal aspects of the root. An example of this proposed method is shown as Figure 5 below. We have also considered ways of automatically correcting errors due to operator selection by specifying a region with geometry instead of points and further extracting a more consistent measure of angle information

4 CONCLUSIONS

Investigating root morphology and the timing of dental development provides additional insight into the factors in the complex adaptive process that gives rise to hypodontia. This study has advanced the techniques available and provides the basis for the further development of this component of the overall study. The idea of automatically correcting operator input to improve reliability of measurements is of general interest.

REFERENCES

[1] Brook, A H. & O'Donnell, M.B., *The dentition: a complex system demonstrating self-* *principles*, 2011 Fifth International Conference on Self Adaptive and Self Organising Systems. C.P.S. Washington IEEE, pp. 208–209, 2011. https://doi.org/10.1109/SASO 2011.41

[2] Brook, A., O'Donnell, M.B., Hone, A., Hart, E., Hughes, T., Smith, R. & Townsend, G., General and craniofacial development are complex adaptive processes influenced by diversity. *Australian Dental Journal*, **59**(S1), pp. 13–22, 2014. https://doi.org/ 10.1111/adj.12158.

[3] Townsend, G., Harris, E.F., Lesot, H., Clauss, F. & Brook, A.H., Morphogenetic fields within the human dentition: a new clinically relevant synthesis of an old concept. *Archives of Oral Biology*, **54**(S1), 34–44, 2009.

[4] Brook, A.H., A unifying aetiological explanation for anomalies of human tooth number and size. *Archives of Oral Biology*, **29**, pp. 373–378, 1984.

[5] Kirkham, J., Kaur, R., Stillman, E.C., Blackwell, P.G., Elcock, C. & Brook A.H., The patterning of hypodontia in a group of young adults in Sheffield. U.K. *Archives of Oral Biology*, **50**, pp. 287–291, 2005.

[6] Brook, A.H. & Holt, R.D., The relationship of crown length to root length in permanent maxillary central incisors. *Proceedings of the British Paedodontic Society*, **8**, pp. 17–20, 1978.

[7] Holt, R.D. & Brook, A.H., Taurodontism: a criterion for diagnosis and its prevalence in mandibular first permanent molars in a sample of 1,115 British schoolchildren. *Journal of the International Association of Dentistry for Children*, **10**(2), pp. 41–47, 1979.

[8] Brook, A.H. & Scheers, M., Variation in tooth root morphology in a Romano-British population. *Dental Anthropology*, **19**(2), pp. 33–38, 2006.

[9] Falk, D., Book Review of 'An introduction to human evolutionary anatomy' by L. Aiello and C. Dean. *Man*, **27**, pp. 410–411. https://doi.org/10.2307/2804064

VARIATIONS IN DENTAL ARCH MORPHOLOGY ARE OUTCOMES OF THE COMPLEX ADAPTIVE SYSTEM ASSOCIATED WITH THE DEVELOPMENTAL VARIATION OF HYPODONTIA

DILAN PATEL[1], SADAF SASSANI[1], MAURO FARELLA[2], SARBIN RANJITKAR[1], ROBIN YONG[1], STEVE SWINDELLS[2] & ALAN BROOK[1, 3]
[1]Adelaide Dental School, the University of Adelaide, Australia.
[2]Faculty of Dentistry, University of Otago, New Zealand.
[3]Institute of Dentistry, Queen Mary University of London, United Kingdom.

ABSTRACT

Development of the human facial structures including the jaws and dentition occurs in a process that has the characteristics of a complex adaptive system (CAS) influenced by epigenetic, genetic and environmental factors. Earlier studies have suggested dental arch development to be reduced in size in subjects with hypodontia when compared with controls. Hypodontia is a variation of development and presents with a reduced number of teeth together with several other phenotypic changes. This study uses enhanced 3D imaging techniques to increase the accuracy of the measurements of dental arches. The sample consists of orthodontic patients, 60 with hypodontia (thirty males and thirty females), and 60 controls matched for age, gender and ethnicity. One operator using an Amann Girrbach Ceramill Map400 3D scanner recorded the 3D images from dental models. The 3D images were then viewed on MeshLab and the accuracy of the measurements were determined through repeat measurement of the same images; this was undertaken with intra- and inter-operator reproducibility. Ten repeat measurements were taken on 10 different models. Validation of the new system was undertaken by repeating the measurements using the standard 2D caliper technique. Arch dimension measurements were determined from distance between the left-hand side first molar to the right-hand side first molar. Similar measurements were also made for the inter-canine width. The results for average intra-operator measurements were 0.33 mm for the maxillary arch and 0.40 mm for the mandibular arch. The difference in average inter-operator reproducibility was also measured for inter-molar arch dimensions at 0.31 and 0.23 mm for maxillary and mandibular arches, respectively. This novel method provides an increased range of measurement of similar accuracy to standard techniques. This study will proceed to establish the variations on the 3D images between the hypodontia subjects and the control group.
Keywords: 3D imaging, dental arch, hypodontia, measurements, morphology, reliability, repeatability.

1 INTRODUCTION

Hypodontia is largely considered to be a variation of dental development with up to 6 congenitally missing teeth excluding third molars. It forms part of the complex adaptive system (CAS) of oro-facial development whereby epithelial tissues and neural crest mesenchyme are involved in delicate signalling pathways with multiple molecular events that overlap each other [1–4]. For a system to be considered as complex and adaptive, it should exhibit the principles of self-organisation and self-adaptation. The features of self-organisation and emergence occur during dental development at the initiation and morphogenetic stages. Separate genetic pathways and genetic/epigenetic/environmental interactions occur simultaneously throughout the dental developmental process; this is evidence of multitasking. The intricate and multi-level processes that occur during odontogenesis demonstrate the characteristics of a CAS [1, 5–8].

© 2018 WIT Press, www.witpress.com
DOI: 10.2495/DNE-V13-N1-107-113

Hypodontia is widely reported to be the most common craniofacial malformation in humans with varying reports of between 2% and 10% of the global population in the permanent dentition excluding third molars [8–11].

The dental arch consists of the dentition and the surrounding structures; alveolar bone forms a large component of the tissues surrounding the dentition. Development of the alveolus is influenced by the presence of teeth. This forms a part of the CAS; however, there is a marked environmental influence during development and function [5, 7, 8].

2 AIM

Dental plaster study models of the maxillary and mandibular dentition have for several decades been used by dental practitioners as a vital aid in treatment planning and form part of the patients' clinical and medico-legal records. Plaster models are especially relevant for orthodontic patients, and many patients with hypodontia require orthodontic treatment as part of the multidisciplinary treatment planning. Plaster models are disadvantageous in that they are relatively delicate and require careful storage and labelling. When many models are collected over a long period, they can take up a considerable amount of space which also incurs a cost. Additionally, when measuring and using models over a long period they are subjected to a degree of wear and tear, which can alter the measurements [12–16].

Three-dimensional scanning has been developed as a recent method for measuring and assessing dental models. With the development of scanning technologies, it has become increasingly relevant to apply this technology to the dental arch and test the accuracy of measuring dental scans between operators [14].

The aim of this paper is to determine the accuracy of 3D scanning technologies in assessing arch dimensions as part of a multicentre, multidisciplinary collaborative study of hypodontia relating genotype and phenotype.

3 MATERIALS AND METHODOLOGY

This study has a unique sample group with defined parameters. The 120 subjects are orthodontic patients from the University of Otago. The hypodontia group was matched for age, gender and ethnicity with equivalent controls. Thirty male hypodontia subjects are in the age range of 10.2–21.0 years old and matched with 30 controls in the age range of 10.1–17.3 years old (Table 1). Thirty female hypodontia subjects are in the age range of 10.8–19.3 matched with controls in the age range of 10.6–19.2 (Table 1). This is summarised in Table 1. The ethnicity of the sample was self-identified by each of the subjects who were matched together. A statistically relevant number of hypodontia and control subjects were recorded [15].

Duplicate die stone models were poured and trimmed from existing models of the maxillary and mandibular dental arches of each subject in the study. Each model was placed in a fixed repeatable position in an Amann Girrbach Ceramill Map400 3D scanner, with the scanner set on its highest detection limit of 20 microns.

Table 1: Study sample of hypodontia patients matched with controls for gender and age.

	Age range of subjects (in years)	
Number in each gender	60 Males	60 Females
Hypodontia	10.2–21.0	10.8–19.3
Controls	10.1–17.3	10.6–19.2

3.1 Validation

Caliper measurements of the same point-to-point measurements were recorded on the dental models using a Craftright 150-mm stainless steel digital Vernier caliper to validate the 3D scans.

3.2 Measurements

To assess the molar width of the arches, measurements of the 3D scans were recorded on the maxillary arch from the disto-buccal cusp tip of tooth 16 to the corresponding cusp on tooth 26 (Fig. 1). Similarly, on the mandibular arch, the distance was measured between the disto-buccal cusp of tooth 36 and the disto-buccal cusp of tooth 46 (Fig. 2). Ten measurements of each arch were recorded from the control group of this preliminary study.

The MeshLab software program was used to record measurements; it is a relatively simple mesh processing system and is often used in processing meshes from 3D scans. The software was chosen for its ease of use in recording simple point-to-point measurements including the ability to rotate the model in any direction, while taking a measurement, this assists in ensuring the highest point of the cusp tip is recorded.

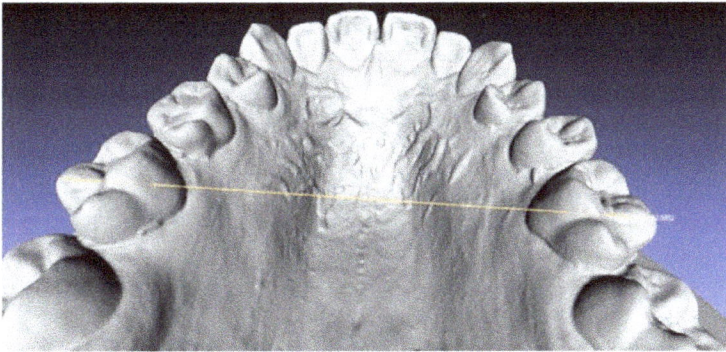

Figure 1: Maxillary inter-molar measurement, magnified view.

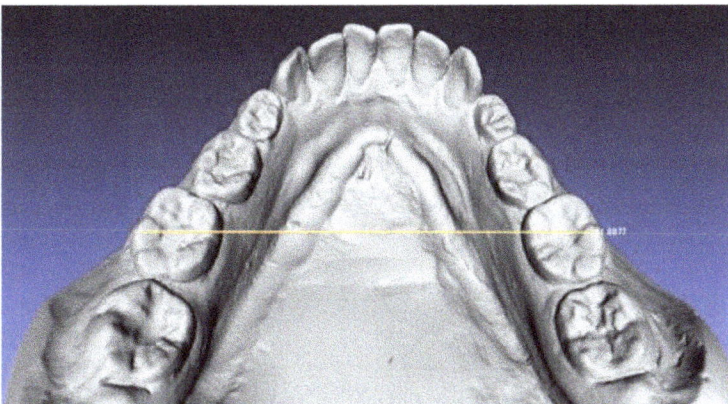

Figure 2: Mandibular inter-molar measurement, full arch view.

3.3 Error determination and repeatability testing

Variability in the measurement process results in multiple repeat measurements of the same distance. In instances where the molar cusp tip was not clearly defined or there was significant wear of the cusp, the reference points for the linear measurement may cause greater variance. Technical error of measurement (TEM) should be considered to determine if a statistically significant difference between repeat measurements exists. To quantify the TEM, a two-way random effect statistical model with single measure and absolute agreement was applied to the data. An intraclass correlation coefficient (ICC) model was used to calculate the above measurements using SPSS software for statistical analysis. To test for intra-operator error, these measurements were repeated 8 weeks later by the same operator (Operator 1). Additionally, a second operator with equal experience in using the MeshLab software (Operator 2) performed the same measurements to test for inter-operator error [15, 16].

4 RESULTS

The results for intra and inter-operator error are given in Tables 2–5 below. All measurements listed are graded in millimetres (mm).

Table 2: Intra-operator error in the maxillary arch (in mm).

Measurement time 1	Measurement time 2	Differential
51.38	51.54	0.16
53.06	53.63	0.57
58.19	58.54	0.35
49.14	49.75	0.61
57.72	57.89	0.17
48.84	49.59	0.75
56.21	56.01	0.20
54.34	54.53	0.19
51.88	51.66	0.22
45.02	44.98	0.04
		Mean: 0.33

Table 3: Intra-operator error in the mandibular arch (in mm).

Measurement time 1	Measurement time 2	Differential
54.82	54.91	0.09
46.69	46.61	0.08
53.06	52.52	0.54
43.17	44.35	1.18
51.18	51.39	0.21
42.89	43.62	0.73
49.26	48.88	0.38
47.99	48.45	0.46
45.13	45.49	0.36
40.21	40.20	0.01
		Mean: 0.40

Table 4: Inter-operator error in the maxillary arch (in mm).

Operator 1	Operator 2	Differential
51.99	51.54	0.45
53.29	53.63	0.34
57.92	58.54	0.62
49.45	49.76	0.31
57.83	57.89	0.06
49.55	49.60	0.05
56.48	56.01	0.47
54.29	54.53	0.24
51.66	51.66	0.00
45.55	44.98	0.57
		Mean: 0.01

Table 5: Inter-operator error in the mandibular arch (in mm).

Operator 1	Operator 2	Differential
54.92	54.85	0.07
46.61	46.84	0.23
52.52	52.30	0.22
44.35	43.71	0.64
51.40	50.91	0.49
43.62	43.55	0.07
48.88	48.91	0.03
48.45	48.07	0.38
45.49	45.36	0.13
40.20	40.21	0.01
		Mean: 0.227

Using the statistical applications mentioned above, the ICC for the maxillary and mandibular arches are as follows:

ICC values	Intra-operator	Inter-operator
Maxilla	0.996	0.996
Mandible	0.993	0.998

5 DISCUSSION

The above results demonstrate the high correlations for both intra-operator and inter-operator dimensions of inter-molar distances. An ICC reading of greater than 0.75 is indicative of good reliability and an ICC of 1 is an identical match in measurements. Studies have also indicated that digital models are a more reliable and clinically acceptable method of assessing inter-molar and inter-canine distances [5, 13–16].

In the literature, it is suggested that variations can arise in measurements relating to the operator position and point chosen on the digitized casts. However, as in our study, the variation is often less than that of caliper hand measurements [13].

6 CONCLUSION

The above results have shown that three-dimensional scanning techniques are an appropriate method of assessing dental arch morphology than using digital calipers to measure dental arch casts, which provides more extensive measurements. The preliminary results of this study demonstrate that it is viable for future detailed analysis of the dental arch morphology in this sample.

ACKNOWLEDGEMENTS

We are grateful for the participation of the patients in this study, we would also like to acknowledge the assistance and advice of Dr M Sassani and Dr D Haag. We arc also grateful to the University of Otago dental laboratory staff for their patience and assistance with this study.

REFERENCES

[1] Brook, A.H., O'Donnell, M.B., Hone, A., Hart, E., Hughes, T.E., Smith, R.N. & Townsend, G.C., General and craniofacial development are complex adaptive processes influenced by diversity. *Australian Dental Journal*, **59**, pp. 13–22, 2014. https://doi.org/10.1111/adj.12158

[2] Wang, J., Sun, K., Shen, Y., Xu, Y., Xie, J., Huang, R., Zhang, Y., Xu, C., Zhang, X., Wang, R. & Lin, Y., DNA methylation is critical for tooth agenesis: implications for sporadic non-syndromic anodontia and hypodontia. *Scientific Reports*, **6**, 19162, 2016. https://doi.org/10.1038/srep19162.

[3] Al Shahrani, I., Togoo, R.A. & AlQarni, M.A., A review of hypodontia: Classification, prevalence, etiology, associated anomalies, clinical implications and treatment options. *World Journal of Dentistry*, **4**(2), pp. 117–125, 2013. https://doi.org/10.5005/jp-journals-10015-1216

[4] Thesleff I., Current understanding of the process of tooth formation: transfer from the laboratory to the clinic. *Australian Dental Journal*, **59**, pp. 48–54, 2014. https://doi.org/10.1111/adj.12102

[5] Brook, A.H., Koh, K. & Toh, V., Influences in a biological complex adaptive system: environmental stress affects dental development in a group of Romano-Britons. *International Journal of Design & Nature and Ecodynamics*, **11**(1), pp. 33–40, 2016. https://doi.org/10.2495/dne-v11-n1-33-40

[6] Koh, K., Toh, V., Brook O'Donnell, M., Ranjitkar, S. & Brook, A.H., A complex adaptive system in which environmental stress affects gene expression during development. *International Journal of Design & Nature and Ecodynamics*, **11**(4) pp. 686–695, 2016. https://doi.org/10.2495/dne-v11-n4-686-695

[7] Lam, F., Yong, R., Ranjitkar, S., Townsend, G.C. & Brook, A.H., Agents within a developmental complex adaptive system: intrauterine male hormones influence human tooth size and shape. *International Journal of Design & Nature and Ecodynamics*, **11**(4), pp. 696–702, 2016. https://doi.org/10.2495/dne-v11-n4-696-702

[8] Brook A.H., Jernvall J., Smith R.N., Hughes T.E. &Townsend G.C., The dentition: the outcomes of morphogenesis leading to variations of tooth number, size and shape. *Australian Dental Journal*, **59**(S1), pp. 131–142, 2014.
https://doi.org/10.1111/adj.12160

[9] Polder, B.J., Van't Hof, M.A., Van der Linden, F.P. & Kuijpers-Jagtman A.M., A meta-analysis of the prevalence of dental agenesis of permanent teeth. *Community Dentistry and Oral Epidemiology*, **32**(3), pp. 217–226, 2004.
https://doi.org/10.1111/j.1600-0528.2004.00158.x

[10] Brook, A.H., Multilevel complex interactions between genetic, epigenetic and environmental factors in the aetiology of anomalies of dental development. *Archives Oral Biology*, **54**S, pp. S3–S17, 2009.
https://doi.org/10.1016/j.archoralbio.2009.09.005

[11] Khalaf K., Miskelly J., Voge E. & Macfarlane T.V., Prevalence of hypodontia and associated factors: a systematic review and meta-analysis. *Journal of Orthodontics*, **41**(4), pp. 299–316, 2014.
https://doi.org/10.1179/1465313314Y.0000000116

[12] Stevens, D., Mir, C., Nebbe, B., Raboud, D., Heo, G. & Major, P., Validity, reliability, and reproducibility of plaster vs digital study models: Comparison of peer assessment rating and Bolton analysis and their constituent measurements. *American Journal Orthodontics and Dentofacial Orthopaedics*, **129**, pp. 794–803, 2006.
https://doi.org/ 10.1016/j.ajodo.2004.08.023

[13] Bell, A., Ayoub, A. & Siebert, P., Assessment of the accuracy of a three- dimensional imaging system for archiving dental study models. *Journal of Orthodontics*, **30**, pp. 219–223, 2003.
https://doi.org/10.1093/ortho/30.3.219

[14] Moreira, D., Gribel, B., Torres, G., Vasconcelos, K., Freitas, D. & Ambrosano, G., Reliability of measurements on virtual models obtained from scanning of impressions and conventional plaster models. *Brazilian Journal Oral Science*, **13**(4), pp. 297–302, 2014.
https://doi.org/10.1590/1677-3225v13n4a11

[15] Harris, E.F. & Smith, R.N., Accounting for measurement error: a critical but often overlooked process. *Archives of Oral Biology*, **54**(1), pp. 107–117, 2008.
https://doi.org/10.1016/j.archoralbio.2008.04.010

[16] Cicchetti, D., Guidelines, criteria, and rules of thumb for evaluating normed and standardized assessment instruments in psychology. *Psychological Assessment*, **6**(4), pp. 284–290, 1994.
https://doi.org/10.1037//1040-3590.6.4.284

VARIATION IN TOOTH CROWN SIZE AND SHAPE ARE OUTCOMES OF THE COMPLEX ADAPTIVE SYSTEM ASSOCIATED WITH THE TOOTH NUMBER VARIATION OF HYPODONTIA

SADAF SASSANI[1], DILAN PATEL[1], MAURO FARELLA[2], MACIEJ HENNEBERG[1], SARBIN RANJITKAR[1], ROBIN YONG[1], STEPHEN SWINDELLS[2] & ALAN H. BROOK[1,3]
[1]The University of Adelaide, the School of Dentistry, South Australia.
[2]The University of Otago, the School of Dentistry, New Zeland.
[3]Queen Mary University, Institute of Dentistry, London.

ABSTRACT

The development of the dentition is a good model of general development; it has the general characteristics of a complex adaptive system. The developmental variation of hypodontia presents with a reduced number of teeth with several other phenotypic changes. The teeth formed are smaller in size, have different crown and root morphology and are delayed in development. The present study is a component of a multi-centre and multidisciplinary collaborative study to investigate hypodontia from genotype to phenotype. This study uses enhanced 3D-imaging techniques in order to increase the range of parameters of the phenotypic outcome: tooth size and tooth shape. The sample consists of orthodontic patients, 60 with hypodontia (30 males and 30 females), and 60 controls matched for age, sex and ethnicity. The material studied for these measurements are the dental models of each patient; these have been imaged with an Amann Girrbach Ceramill Map400 3D scanner. The 3D images produced were all taken by one operator and viewed on MeshLab. The accuracy of the measurements taken was determined through repeat measurements of the same images, undertaken to determine intra and inter-operator reproducibility. This new system was validated by repeating these measurements using the standard 2D caliper technique. Ten repeat measurements were taken on ten models of the lower and upper premolar inter-cuspal distances. The average intra-operator reproducibility for the inter-cuspal distances when measuring the distance between the buccal and palatal cusp of the maxillary premolar was 0.20 mm; the mandibular premolar was 0.32 mm. The results for inter-operator reproducibility demonstrate an average difference of 0.24 mm for the maxillary premolar and 0.16 mm for the lower premolar. This novel method provides an increased range of measurements with good levels of accuracy. This study will go on to establish the variations on the 3D images between the hypodontia and the control group.
Keywords: 3D-Imaging, complex adaptive system, error, hypodontia, inter, intra, linear, measurement, reliability, reproducibility.

1 INTRODUCTION

This is one component of a multi-centre and multi-disciplinary study to investigate the complex relationship between the genotype and phenotype in hypodontia patients. The component reported here will compare the crown dimensions in patients with hypodontia to matched controls. Furthermore, as novel 3D measurement techniques are used it is necessary to determine the intra and inter-operator error to ascertain the degree of operator reliability; this paper will outline the procedure used to determine this.

2 HYPODONTIA

The majority of the population develops 20 primary teeth and 32 secondary teeth. When a specific tooth does not form this is known as tooth agenesis; the most frequently affected teeth are the third molars [1]. The dentition may be affected in varying numbers of missing teeth; hypodontia, which is the absence of one or up to five teeth, oligodontia, which involves six teeth or

© 2018 WIT Press, www.witpress.com
DOI: 10.2495/DNE-V13-N1-114-120

more and anodontia, which is the total absence of teeth [1]. Oligodontia and anodontia occur infrequently in a population and are commonly associated with syndromes such as ectodermal dysplasia [2]. However, hypodontia is one of the most commonly occurring human dental variations [3]. The remaining teeth present in an individual affected by hypodontia are smaller in dimensions, have different crown and root morphology and are delayed developmentally.

In biological development complex adaptive systems (CAS) refer to a dynamic process where lower level interactions produce higher level structure and phenomena [4]. Dental development has been shown to have the features of a CAS and can provide a good paradigm for general development as well as links with other autonomous complex adaptive systems such as fingerprint development [4–6]. Key characteristics and components of dental development include multidimensional, multi-level and multifactorial properties; the resulting phenotype is the product of multiple interactions between epigenetic, genetic and environmental factors [4–7]. Environmental changes within the developing dental CAS, such as increased intrauterine male hormones, have demonstrated phenotypic changes in tooth size, shape and dental arches [8, 9].

3 MATERIALS AND METHODS

3.1 Study sample

The sample comprises 30 female and 30 male orthodontic patients from The University of Otago with mild-to-moderate hypodontia (one to five missing teeth) in their permanent dentition. This has been confirmed via an orthopantomograph. A corresponding group of 30 female and 30 male patients who have a complete dentition acts as the controls. The age range of the male hypodontia group was 10.2–21.0 years, the female hypodontia group was 10.8–19.3 years, the male control group was 10.1–17.3 years and the female control group was 10.6–19.2 years. The sample size was estimated by Brook *et al.* [10]; suggesting a comparison of two groups of twenty will give an 80% power to determine a size difference of 0.90 millimeters. The samples used to determine the intra- and inter-operator error were from ten additional patients that were a part of the control pool.

3.2 Study materials

The patients' dental casts were duplicated using polyvinyl siloxane (PVS) impression material at The University of Otago. The impressions were poured in die stone and trimmed. The dental casts were then scanned using an Amann Girrbach Ceramill Map400 by the same operator; this produced a 3D digital model in a STL file format.

3.3 Ethical approval

Ethical approval was granted by the Human Research Ethics Committee of The University of Adelaide; this work was deemed to be of negligible risk.

3.4 3D-Image analysis

The 3D digital model, or images, produced from the dental casts were extremely detailed and data dense, in part because they were constructed from a twenty-micron resolution or less. The 3D models were then measured on MeshLab software for the premolar inter-cuspal distance.

3.5 Inter-cuspal distance

The teeth selected for measurement were the upper second premolar on the right side, and the lower second premolar on the left side. The inter-cuspal distance was defined as the distance between the cusp tips (Fig. 1). Some lower premolars have additional cusp tips, in this instance it was decided to take the distance from the distal cusp tip.

3.6 Intra- and inter-operator error

The intra-operator error was determined by having the first operator take ten initial inter-cuspal measurements from the upper and lower second premolar. The first operator then took the same measurements from the same teeth eight weeks later and the differentials between the two sets of measurements were calculated.

The inter-operator error was determined by having a second operator take the same ten inter-cuspal measurements from the upper and lower second premolar, the differentials between the first and second operator were then calculated.

4 RESULTS

The results are presented in the following Tables 1–4. Tables 1 and 2 show the values the first operator measured on two separate occasions with the differentials (intra-operator), followed by the mean of the differentials, which were 0.20 mm and 0.32 mm for teeth the upper and lower second premolar, respectively. Tables 3 and 4 show the values the first operator and the second operator measured followed by the difference between them (inter-operator); the average for this was 0.09 mm and 0.58 mm for the upper and lower second premolar, respectively. Using SPSS software an intra-class correlation coefficient model (ICC), specifically a two-way random effects with absolute agreement, was applied to the values produced by the first and second operators to assess their reliability. Values produced for an ICC range from 0 to –1, where 1 is a perfect correlation. The ICC for the first operator's intra-operator reliability was 0.818 for the upper second premolar and 0.852 for the lower second premolar. The ICC for the first operator's and the second operator's inter-operator reliability was 0.955 for the upper second premolar and 0.685 for the lower second premolar.

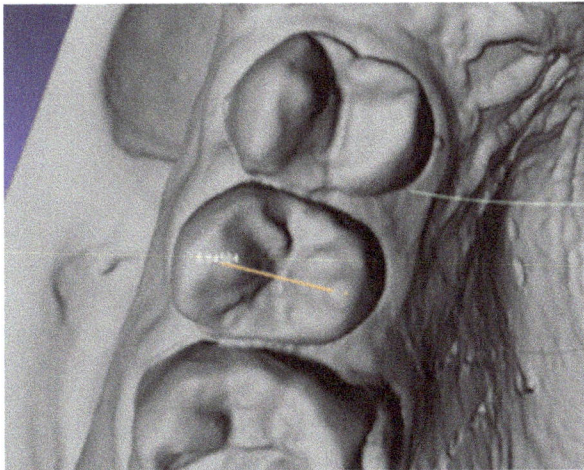

Figure 1: Premolar inter-cuspal distance measurement.

Table 1: First operator intra-operator error differentials for the upper second premolar (mm).

First operator measurements of the upper second premolar	First operator measurements of the upper second premolar eight weeks later	Differentials
5.49148	5.72686	0.23538
5.7727	5.5659	0.2068
5.89043	5.61694	0.27349
n/a	n/a	n/a
6.2691	6.12684	0.14226
n/a	n/a	n/a
5.4855	5.35244	0.13306
6.40191	6.20258	0.19933
5.50001	5.72432	0.22431
n/a	n/a	n/a

Mean: 0.20209

Table 2: First operator intra-operator error differentials for the lower second premolar (mm).

First operator measurements of the lower second premolar	First operator measurements of the upper second premolar eight weeks later	Differentials
5.96405	6.09306	0.12901
4.08941	4.53681	0.4474
5.87754	6.02671	0.14917
n/a	n/a	n/a
5.28945	5.47873	0.18928
4.30977	3.26884	1.04093
5.17975	5.07707	0.10268
5.45517	5.22245	0.23272
5.0966	4.85478	0.24182
n/a	n/a	n/a

Mean: 0.31662625

5 DISCUSSION

ICC values of above 0.75 are considered to indicate good reliability [11, 12]. All the values for intra- and inter-operator reliability, except for the lower second premolar, were above this threshold and it suggests there is evidence of reliability. The reasons for the lower second premolar having decreased reliability likely stems from its morphology; it can have one or two lingual cusp tips and this could be open to interpretation by the operator. Some degree of variability between operators will always exist when operators choose anatomical points because of their interpretation of where that anatomical landmark is. Variability also exists between operators measuring with traditional calipers, plaster models, and where the points have been

Table 3: First and second operator inter-operator error for the upper second premolar (mm).

First operator upper second premolar measurements	Second operator upper second premolar measurements	Differentials
5.49148	5.538	0.04652
5.7727	5.709	0.0637
5.89043	6.089	0.19857
n/a	n/a	n/a
6.2691	6.34	0.0709
n/a	n/a	n/a
5.4855	5.48	0.0055
6.40191	6.233	0.16891
5.50001	5.583	0.08299
n/a	n/a	n/a

Mean: 0.091012857

Table 4: First and second operator inter-operator error for the lower second premolar (mm).

First operator lower second premolar measurements	Second operator lower second premolar measurements	Differentials
5.96405	5.63	0.33405
4.08941	3.77	0.31941
5.87754	5.377	0.50054
n/a	n/a	n/a
5.28945	5.127	0.16245
4.30977	3.325	0.98477
5.17975	3.969	1.21075
5.45517	5.212	0.24317
5.0966	4.195	0.9016
n/a	n/a	n/a

Mean: 0.5820925

marked for them to measure; simply due to the slight variation in the manual positioning of the calipers [13]. Previous studies have demonstrated that, while there is inter-operator error when measuring on a 3D model, this is still less than traditional plaster models and calipers [13–15]. Digital models may be more reliable than hand calipers because of the ability to zoom in and rotate the models in multiple planes [15].

6 CONCLUSION

The evidence from this study and the literature suggests that measurements taken on 3D digital models are more reliable than traditional caliper hand measurements. In the future if all dental models are stored as 3D-images, with a high quality and accurate resolution, the primary source of error amongst operators will stem from interpretation of anatomical landmarks. The initial study reported here provides a basis for the future detailed studies of the crown size and shape in these hypodontia patients and controls.

ACKNOWLEDGEMENTS
We are grateful for the participation of the patients in this study and we would also like to acknowledge the assistance of Dr. M Sassani, Dr. D Haag, and the work of the dental technicians at The University of Otago.

REFERENCES

[1] AlShahrani, I., Togoo, R.A. & AlQarni, M.A., A review of hypodontia: classification, prevalence, etiology, associated anomalies, clinical implications and treatment options. *World Journal of Dentistry*, **4**(2), pp. 117–125, 2013.
https://doi.org/10.5005/jp-journals-10015-1216

[2] Werther, R. & Rothenberger, F., Anodontia, a review of its etiology with presentation of a case. *American Journal of Oral Surgery*, **25**, pp. 61–81, 1939.
https://doi.org/10.1016/S0096-6347(39)90349-2

[3] Brook, A.H., Dental anomalies of number, form and size: their prevalence in British schoolchildren. *Journal of the Internatioanl Association of Dentisity for Children*, **5**(2), pp. 37–53, 1974.

[4] Brook, A.H., O'Donnell, M.B., Hone, A., Hart, E., Hughes, T.E., Smith, R.N. & Townsend, G.C., General and craniofacial development are complex adaptive processes influenced by diversity. *Australian Dental Journal*, **59**, pp. 13–22, 2014.
https://doi.org/10.1111/adj.12158

[5] Taduran, R.J.O., Ranjitkar, S., Hughes, T., Townsend, G. & Brook, A.H., Complex systems in human development: sexual dimorphism in teeth and fingerprints of Australian twins. *International Journal of Design & Nature and Ecodynamics*, **11**(4), pp. 676–685, 2016.
https://doi.org/10.2495/DNE-V11-N4-676-685

[6] Brook, A., Koh, K. & Toh, V., Influences in a biological complex adaptive system: environmental stress affects dental development in a group of Romano-Britons. *International Journal of Design & Nature and Ecodynamics*, **11**(1), pp. 33–40, 2016.
https://doi.org/10.2495/DNE-V11-N1-33-40

[7] Koh, K., Toh, V., Brook O'Donnell, M., Ranjitkar, S. & Brook, A., A complex adaptive system in which environmental stress affects gene expression during development. *International Journal of Design & Nature and Ecodynamics*, **11**(4), pp. 686–695, 2016.
https://doi.org/10.2495/DNE-V11-N4-686-695

[8] Lam, F., Yong, R., Ranjitkar, S., Townsend, G.C. & Brook, A.H., Agents within a developmental complex adaptive system: intrauterine male hormones influence human tooth size and shape. *International Journal of Design & Nature and Ecodynamics*, **11**(4), pp. 696–702, 2016.
https://doi.org/10.2495/DNE-V11-N4-696-702

[9] Patel, P., Yong, R., Ranjitkar, S., Townsend, G. & Brook, A., Agents within a development complex adaptive system: intrauterine male hormones and dental arch size in humans. *International Journal of Design & Nature and Ecodynamics*, **11**(4), pp. 703–711, 2016.
https://doi.org/10.2495/DNE-V11-N4-703-711

[10] Brook, A.H., Elcock, C., Al-Sharood, M.H., McKeown, H.F., Khalaf, K. & Smith, R.N., Further studies of a model for the etiology of anomalies of tooth number and size in humans. *Connective Tissue Research*, **43**(2–3), pp. 289–295, 2002.
https://doi.org/10.1080/03008200290000718

[11] Cicchetti, D., Guidelines, criteria, and rules of thumb for evaluating normed and standardized assessment instruments in psychology. *Psychological Assessment*, **6**(4), pp. 284–290, 1994.
https://doi.org/10.1037/1040-3590.6.4.284

[12] Harris, E.F. & Smith, R.N., Accounting for measurement error: a critical but often overlooked process. *Archives of Oral Biology*, **54**(1), pp. 107–117, 2008.

[13] Bell, A., Ayoub, A. & Siebert, P., Assessment of the accuracy of a three- dimensional imaging system for archiving dental study models. *Journal of Orthodontics*, **30**, pp. 219–223, 2003.
https://doi.org/10.1093/ortho/30.3.219

[14] Moreira, D., Gribel, B., Torres, G., Vasconcelos, K., Freitas, D. & Ambrosano, G., Reliability of measurements on virtual models obtained from scanning of impressions and conventional plaster models. *Brazilian Journal of Oral Sciences*, **13**(4), pp. 297–302, 2014.
https://doi.org/10.1590/1677-3225v13n4a11

[15] Stevens, D., Mir, C., Nebbe, B., Raboud, D., Heo, G. & Major, P., Validity, reliability, and reproducibility of plaster vs digital study models: comparison of peer assessment rating and Bolton analysis and their constituent measurements. *American Journal of Orthodontics and Dentofacial Orthopedics*, **129**, pp. 794–803, 2006.
https://doi.org/10.1016/j.ajodo.2004.08.023

ON THE FORMATION OF MEXICO STATE

D.E. SANTOS REYES
I'chi Research and Engineering, Santos Reyes Yucuna, Huajuapan de Leon, Oaxaca, Mexico.

ABSTRACT

The process of change of very large systems, such as a country, is essentially described as a succession of chronological events. This viewpoint provides some critical issues regarding the development of the system of interest. However, it may not describe the dynamics of critical historical events occurring throughout the formation of the system. There is a need to pay attention not only to critical historical events, but also to the dynamics of the step changes. How did Mexico State evolve to be the way it is today? How regulated has it been through time? This is an attempt to escape from the traditional chronological description of historic events. Complexity science provides new and more fruitful conceptions, and more effective ways for studying self-organisation and self-regulated complex systems. This paper explores a major step change of the origin of modern Mexico State: pre-Aztec and Aztec dominion. The aim is to develop a fundamental understanding of the complexity of Mexico State system behaviour throughout its formation. This is with the aim to develop models that can be employed in engineering the future step changes of the system.

Keywords: change, complexity, novelty, systems.

1 INTRODUCTION

Complexity science studies the emergent behaviour of highly interconnected systems, such as healthcare, education, economic and social systems [1–3]. On the other hand, there has been a significant effort by historians to develop a great deal of understanding of historical events throughout the formation of very large social systems such as a country. However, very little attention has been given to the study of the dynamics of the complex process of formation of such systems through time. This papers explores the following issues: how very large systems, such as a country: Mexico, might have evolved through a very long period of time. This is, how did Mexico State evolved to be the way it is today? How regulated has it been through time? Section two revises relevant issues related to complexity and systems whilst section three presents the approach adopted here to study the situation of interest. Section four describes the initial insights gained at this exploration stage of the study and finally some concluding remarks are provided in section five.

2 COMPLEXITY, SYSTEMS AND STATE FORMATION

2.1 Complexity and systems

According to the online Collins English dictionary [4], complexity is 'the state or quality of being intricate or complex.' Similarly, complexity is defined by scholars as the number of distinguishable states of the elements of a given situation or system and the measure of complexity is called variety [5–7]. Complexity science, on the other hand, is an approach to study the emergent behaviour of complex systems, such as healthcare, education, economic, and social systems [8–10]. A system or a given situation is usually understood as a set of interrelated entities or parts. According to Beer [5], systems are distinguished by a set of interrelated parts, which are a systematically arranged assemblage, and the system serves a clear function or purpose. Furthermore, Beer emphasises that systems are very complex, highly probabilistic, and to some extent self-regulating. The systemic nature of very complex systems provides us with the very features required for the study of their behaviour and control.

2.2 Social systems

Biologists define evolution as a gradual change or development of living systems to a more complex state. Scholars have used this insight to study the behaviour of man-made systems, including social systems [11, 12]. According to some contemporary views, evolution may be characterised by phases of equilibrium interrupted by changes and destruction [13]. Catastrophic events, such as natural disasters, economic crisis, social unrest, etc. can be said to be systems' response to input changes at the verge of chaos [7]. These transient responses of systems affect dramatically on human life and on the natural environment. Such events may be regarded as the consequence of the obtrusive relationships between agents or parts that constitute a given situation or system. It seems that the connectivity of systems controls these novel catastrophic events. This provides great opportunity for scientific understanding of complex systems. The future time evolution of very complex social systems would be inherently very difficult to predict [14]. This does not prevent, however, the application of the scientific method for the study of novel or extreme events, as discussed by many scholars, such Boisot and McKelvey [15]. The notion of complex systems as pervasive in real life situations is relatively new, and its application to develop understanding of the intricate complexity of social systems is increasing. The behaviour of social systems is regulated by the richness of their connectivity and interactions.

2.3 State formation

History may be defined as the study of the past, i.e. the production of knowledge about past events. Marwick [16] emphasises that 'history is the body of knowledge about the past, as produced by historians, and the past is everything which actually happened, whether known, or written about by historians, or not.' In general, past events are studied and reported as a succession of chronological isolated events. State formation from an anthropological view point is to study all human culture and related issues [17]. The issue addressed here is how very large systems, such as a country: Mexico, might have evolved through a very long period of time. This is how did Mexico State evolve to be the way it is today? How regulated has it been through time?

3 THE APPROACH

There has been a significant effort by historians to develop a great deal of understanding of historical events throughout Mexico state formation. However, very little attention has been given to the study of the dynamics of the complex process of the formation of Mexico through time. Thus, a ready-made database does not exist for the study of the evolution of Mexico as a complex system. Figure 1 shows the system-of-focus. Mexico does not exist in isolation, i.e. it is part of the world system. Similarly, it consists of a collection of interrelated subsystems that serve a clear purpose: survival of the whole. A possible benefit of this representation is the ability to shift analysis from the system-of-focus to the lower level, i.e. subsystems level and to the world system. This enables one to analyse not only Mexico as a whole and its connectivity within the world system, but also the connectivity of various subsystem parts of the Mexico State. Furthermore, one can focus on the evolution of the complex

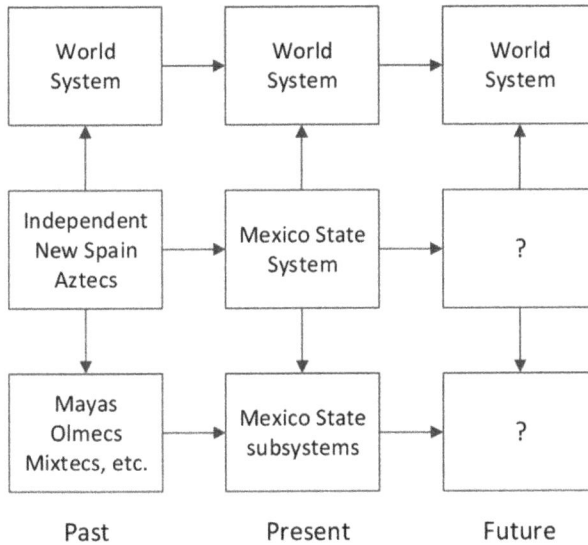

Figure 1: The system of focus: Mexico State System (adopted from [18]).

system through a very long period of time. In this exploratory study, the historian's approach is used in order to develop an understanding about how very large systems, such as Mexico evolved from a series of interactions over a very long period of time. This is, study the past of Mexico by gathering data/facts from archives, books and on-line published records from the first nations, such as Mayas, Olmecs, Zapotecs, Mixtecs, Chichimecs, etc. Finally, with the aid of complexity science principles one should be able to unfold the complexity of the formation of Mexico, from lower to higher levels of recursion and from the distant past to the present, including the near future.

4 THE MEXICO STATE SYSTEM

4.1 The present situation

Mexico's official name is the United Mexican States. It is formed by 31 federal states and its capital city for the past two centuries is called Federal District. It is where the federal executive, legislative, and judicial systems reside. Currently, the Federal District name is in the process of change from a federal district to Mexico City, along with a new Constitution. Mexico is a federal republic with a territory of almost two million square kilometres [19]. Its territory has changed significantly through time. Currently, limits with the United States in the North and with Guatemala and Belize in the Southeast, and with an estimated population of over 120 million inhabitants. Today Mexico is considered as an emergent economy [20]. Its economic stability has been sustained according to the rules of the world-system. Its gross domestic product (GDP) is very poor. In addition to this, the GDP does not consider social welfare and other issues that affect the economy. Mexico is a member of the United Nations, the Organisation of Economic Cooperation and Development (OECD), the World Trade Organisation (WTO) amongst other organisations. The Mexican economic performance has gone through a series of crisis and as a result of this its economic growth is very poor. The mexican socio-economic- and political situation is about surrogate worlds. It is easy enough to accept that everything is steadily improving in the world of the very few (the 20 most powerful Mexican families) whilst the world of millions is getting steadily worse. It is acknowledged by many that the gap between the poor and rich is bigger than ever before [21].

On the other hand, corruption, impunity and crime are endemic. Mexicans have witnessed systemic corruption, which began in the early 1980s. Mexican institutions have automatically transferred public resources to special interests by virtue of formal structures, processes, and paradigms or methods. As discussed elsewhere [22] ordinary crime, and organised crime, such as human and weapon smuggling, drug trafficking, and terrorism have steadily increased over the last decades. Similarly, the healthcare and education systems are mostly funded by public money. However, the education system is characterised by poor performance at all levels. Significant effort has been paid by the Mexican government, over the last decades, to implement a healthcare system that intends to provide a comprehensive health service to a great number of Mexicans [23]. But, this system emphasises on a reactive approach rather than on proactive means to address more affectively the wellbeing of the Mexicans.

4.2 The origins

According to historians, Mesoamerica was a region that covered southern Mexico, Guatemala, Belize, Honduras, and El Salvador. This region was the home of the modern Mexican ancestors, namely Olmecs, Mayas, Zapotecs, Mixtecs, Otomies, Toltecs, Aztecs, amongst others. In 1521, the Spanish conquered the Aztecs and spread their control throughout this region. The Spanish dominion, as viceroyalty of New Spain, extended for about three hundred years. Following the war of independence, (1810–1821), the New Spain became Mexico. Throughout the post-independence period, the newly constituted country went through a series of economic instability and political changes. For example, Mexico lost one-third of its territory to the USA in the Mexican-American war from 1846 to 1848. The stability of the country was further weakened by foreign intervention, such as that in 1862 by France, Great Britain (GB) and Spain. Furthermore, the country was involved in a civil war, and a domestic dictatorship throughout the second half of the 19th century. Finally, the Mexican revolution (1910–1921) led to the formation of the modern Mexico. In line with this, the stages of Mexico State formation can be distinguished as four major step changes namely the Aztec dominion, the Spanish conquest, the independence, and the Mexican revolution. The below section explores the Aztec domain from the perspective of complex system theory.

4.2.1 The Aztec dominion
The formation of Mexico can be traced back to the Mesoamerican period. The beginning consisted primarily in isolated nations (Fig. 1a) namely Olmecs, Mayas, Mixtecs, Toltecs, Zapotecs, Aztecs, etc. Recently, historians have recognised that some of these advanced civilisations appeared before others and some disappeared before others. However, they all existed and interacted at the same time at some point in the Mesoamerican period. Furthermore, the descendants of these civilisations, such as the Mixtecs, Zapotecs and Mayas still exist today. Many still walk barefoot and speak a language of their own. At the early stages of the formation of these nations, there were little or no interactions at all. They did not pertain to a coherent whole, thus the dashed circle in Figure 1a. The complexity, or the variety, of this situation may be defined as the number of distinguishable nations: 5. Nothing more is known about it.

As time passed, these nations developed a sophisticated network of interactions with each other. Initially, the interaction of these nations may be characterised by, according to historians, economic or trade relationships. Later, these interactions became obtrusive and patterned (Figure 1b). This connectivity was characterised not only by trade, but also by social and political relationships. This adds more information to the analysis, i.e. the lines denote the complexity of this set of nations. According to Beer [5], the internal relationships within systems should be capable of change. This is, each relationship should take more than one state. This enables the system to evolve. Considering two states, the complexity of the set of nations in Figure 1b may be defined as $2^{n(n-1)} = 2^{5(5-1)} = 2^{20}$. This is the complexity of the cooperative organisation of the set of nations in Figure 2b. However, this system was not regulated by a centralised control system, i.e. it may be regarded as self-organised system. It was disturbed at a certain period of time by the Aztecs. This may be regarded as one of the first step changes towards the birth of a very large system: Mexico. The Aztecs, after a period of pilgrimage, settled in the central Anahuac Valley, today Mexico City. The Aztecs dominated the Mesoamerican people until 1521. The Aztecs created a centralised control over the original people. Thus the continuous line circle in Figure 1c. However, although these people were under the Aztec's regulation they did not serve a clear purpose to the whole Aztec system.

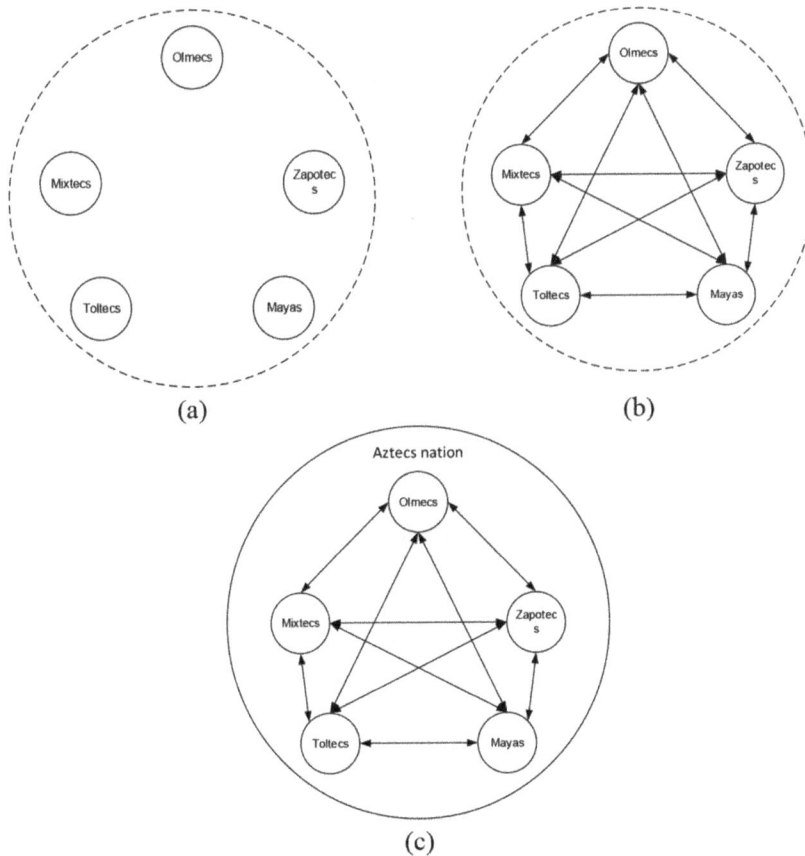

Figure 2: (a) early isolated nations, (b) dynamic system, (c) Aztecs nation.

Owing to the resistance (inertia) of each group of these people, the change or transition to a new situation took a long period, in some cases the new situation, i.e. the Aztecs dominion, was never accepted by many original people. It may be said that the Aztecs dominion was a system of continuous oscillation. There was a need for some methods to bring the oscillations to an end. The whole Aztec system had little or no inherent damping, and the problem of obtaining a suitable response was the Spanish 'conquistadores'.

5 CONCLUSIONS

This paper explored the origins of the formation of Mexico State with the aid of complex system principles. It argued that the original peoples, namely Mayas, Olmecs, Zapotecs, Mixtecs, Toltecs, amongst others, developed a complex assemblage of relationships in the form of trade, social and political issues. However, this complex system did not develop a centralised form of control. It may be said that it was a self-organised system. The evolution of this system was disrupted by the Aztecs who established a centralised control over the original peoples. However, this can be considered as the first step change for the formation of Mexico. Future work will focus on detailed analysis of this initial exploration stage by means of qualitative and quantitative methods. Also, this will include qualitative and quantitative analysis of the other step changes: Spanish conquest, independence, and the so called Mexican revolution.

REFERENCES

[1] Jacobson, M. J. & Wilensky, U., Complex systems in education: scientific and educational importance and implications for the learning sciences. *Journal of the Learning Sciences*, **15**(1), pp. 11–34, 2006.
https://doi.org/10.1207/s15327809jls1501_4

[2] Diez Roux, A.V., Complex systems thinking and current impasses in health disparities research. *American Journal of Public Health*, **101**(9), pp. 1627–1634, 2011.
https://doi.org/10.2105/ajph.test.2011.300149

[3] Sawyer, R.K., *Social emergence: Societies as complex systems*, Cambridge University Press, NY, 2005.

[4] Collins English Dictionary, available at: https://www.collinsdictionary.com/dictionary/english/complexity.

[5] Beer, S., *Decision and Control*, John Wiley & Sons, 1994.

[6] Ashby, W.R., *An introduction to cybernetics*, Chapman & Hall, London, 1957.

[7] Boisot, M. & McKelvey, B., Integrating modernist and postmodernist perspectives on organizations: a complexity science bridge. *Academy of Management Review*, **35**(3), pp. 415–433, 2010.
https://doi.org/10.5465/amr.2010.51142028

[8] McKelvey, B., Salmador, M.P., & Rodriguez-Anton, M., Towards an econophysics view of intellectual capital dynamics: from self-organized critically to the stochastic frontier. *Knowledge Management Research & Practice,* **11**, pp. 142–161, 2013.
https://doi.org/10.1057/kmrp.2013.18

[9] Helbing, D., Managing complexity in socio-economic systems. *European Review*, **17**(2), pp. 423–438, 2009.
https://doi.org/10.1017/s1062798709000775

[10] Davis, B. & Simmt, E., Understanding learning systems: mathematics education and complexity science. *Journal for Research in Mathematics Education*, **34**(2), pp. 137–167, 2003.
https://doi.org/10.2307/30034903

[11] Holland, J.H., Studying complex adaptive systems. *Journal of Systems Science and Complexity*, **19**, pp. 1–8, 2006.
https://doi.org/10.1007/s11424-006-0001-z

[12] Lansing, J.S., Complex adaptive systems. *Annual Review of Anthropology*, **32**, pp. 183–204, 2003.

[13] Turcotte, D.L. & Rundle, J.B., Self-organized complexity in the physical, biological, and social sciences. *PNAS*, **99**(suppl. 1), pp. 2463–2465, 2002.

[14] Health Care as a Complex Adaptive System: Implications for Design Management. *The Bridge*, Spring, pp. 17–25, 2008.

[15] Boisot, M. & McKelvey, B., Connectivity, extremes, and adaptation: a power-law perspective of organizational effectiveness. *Journal of Management Inquiry*, **20**(2), pp. 119–133, 2011.
https://doi.org/10.1177/1056492610385564

[16] Marwick, A., *The fundamentals of history*, available at: https://www.history.ac.uk/ihr/Focus/Whatishistory/marwick1.html

[17] Alonso, A.M., The politics of space, time and substance: State formation, nationalism, and enthinicity. *Annual Review of Anthropology*, **23**, pp. 379–405, 1994.
https://doi.org/10.1146/annurev.anthro.23.1.379

[18] Altshuller, G.S., *Creativity as an exact science: the theory of the solution of inventive problems*, Amsterdam: Gordon and Breach, 1984.

[19] INEGI: Referencias geograficas y extension territorial de Mexico, available at: http://www.inegi.org.mx/inegi/SPC/doc/internet/1-GeografiaDeMexico/MAN_REFGEOG_EXTTERR_VS_ENERO_30_2088.pdf

[20] Aulakh, P.S., Kotabe, M. & Teegen, H., Export strategies and performance of firms from emerging economies: evidence from Brazil, Chile, and Mexico, *Academy of Management Journal*, **43**(3), pp. 342–361, 2000.
https://doi.org/10.2307/1556399

[21] Frausto, S., *Los doce mexicanos mas pobres: el lado B de la lista de millonarios*, Planeta, 2016.

[22] Santos-Reyes, D.E. & Santos-Reyes, J.R. Patterns of temporal diffusion of crime in Mexico. *International Journal of Safety and Security Engineering*, **2**(1), pp. 54–68, 2012.
https://doi.org/10.2495/safe-v2-n1-54-68

[23] Knaul, F.M. & Frenk, J., Health insurance in Mexico: achieving universal coverage through structural reform *Health Affairs*, **24**(6), pp. 1467–1476, 2005.
https://doi.org/10.1377/hlthaff.24.6.1467

GROUP MODELING BUILDING: HOW ENVIRONMENT, CULTURE AND WORK CONDITIONS IMPACT ON THE PROCESS

RINA SADIA
Lecturer – Shenkar College of Engineering and Design, Israel.

ABSTRACT

The objective of this paper is to study the impact of culture and work conditions on the process of group modeling building. This process took place in an Israeli factory, in a country of mixture cultures and social backgrounds. The process of building a model involved many participants of different positions in the factory, composing a diverse group with varied inputs. Since the participants were chosen from various levels within the company, they were also from diverse backgrounds in terms of their cultural background, socio-economic status and their work position. These impact their way of thinking and their opinions on the problem. This research applied the existing techniques of group model building process in the Israeli factory, showing that the implementation of this process in real-life differs from theory and requires additional information and tools. This need rises mainly from the dissimilarity of the cultural and social backgrounds of the organization and the workers, differences in the educational level of the employees and their various occupational statuses. These distinct differences suggest revisions and additions to the way this process is performed, such as the need to improve communication skills between participants, the need to establish rules-of-conduct in large groups with diverse backgrounds, and the value of personal conversations in addition to the group process. It is therefore vital for research management teams to acknowledge these differences between group members in order to understand the contradictory information that may come up from different parts of the group during the model-building process, as well as improve the final outcome of the group model building process. Revealing this kind of cultural mixture allows a continuous improvement process of knowledge elicitation through this model building process, thus improving the work of research management teams.
Keywords: culture, feedback loops, group modeling process, knowledge elicitation.

1 INTRODUCTION

The systems approach distinguishes itself from the more traditional analytical approaches by emphasizing the interactions and connections between the different components of a system. The interactions of the parts become more relevant to understanding the system than understanding the parts. This is the value of systems theory – the whole is more than the sum of its parts. A system is called complex due to the multiplicity of its elements, whether they are natural, technical, economic or social systems. A complex system can be characterized by the interactions between the various parts of the system, which cause internal complex causal structure subjected to feedbacks, and by delayed behavioral reactions, which are counterintuitive and difficult to predict [1].

System dynamics is a methodology, which deals with complex systems. Systems dynamic looks at the same type of problems using the same perspective as in systems thinking. The two fields share the same causal loop mapping techniques, but system dynamics takes an additional step by constructing computer simulation models to confirm that the hypothetical structure can lead to the observed behavior and to test the effects of alternative policies on key variables over time. There are two types of feedback loops: positive loop, which generates a reinforcing and exponential behavior over time, and negative loop, which generates a balancing and equilibrating behavior over time. Interactions between these two types of loops create complex system behaviors of growth and collapse, oscillations and others [1]. The

© 2018 WIT Press, www.witpress.com
DOI: 10.2495/DNE-V13-N1-128-135

modeling process using system dynamics is usually carried out through a group modeling process.

The group modeling process, which is an important process of system dynamics intervention in organizations, is mostly developed by researchers from similar backgrounds and cultures. A research conducted in Israel, a country with a mixture of cultures and social backgrounds, evokes a different experience and approach to the more known group model building (GMB) techniques. The GMB process described above was a part of a larger research exploring the relations between organizational effectiveness, quality culture and employee's health. In order to come up with new insights about GMB, the popular principles of group modeling are reviewed and the firm in which this research took place is subsequently described briefly. In the end, the insights regarding the GMB process and possible improvements are presented, including some modeling process learning critique and conclusions.

The system dynamics group modeling process is primarily based on three fundamental tasks: the first is to elicit knowledge and to reveal the participants' mental models through group participants' interactions. Modelers must elicit knowledge from those that are involved in both operations and/or provide policy decisions in order to develop a useful model that has credibility in the eyes of the managers. This information is then used to develop the model [2, 3]. The second task requires the use of the system dynamics modeling approach to create the conceptual model of the problem by focusing on key feedback loops. The third task includes the conversion of the conceptual model through mathematical formulations to a computer simulation model that allows for the representation of the behavior over time as well as for policy analysis [2, 3]. While each of these three tasks is important and essential for the modeling process, the first task is paramount in determining the quality level of the entire modeling process. The entire group modeling process cannot be better than the information gathered during the initial group sessions. The outcomes of the first task influence the subsequent group modeling processes, and therefore, influence their outcome's quality. This paper deals especially with the first task.

Different situations, cultures and organizational histories can lead to different experiences and needs concerning the application of the GMB process [4]. Different types of interventions exhibit feasibility variation and emphasize the importance of knowing and learning from the wide variety approaches that were used to involve the client in the model building process [5]. Sharing the differences stemming from the differences among situations, backgrounds, cultures and organizational histories is significant for the process of learning and gaining a better insight for a more general effective GMB process. A certain case study, in which systems dynamic was used, provided insights and raised issues concerning the GMB process application [6]. Hereby, we will overview the principles of GMB from the literature, its application to the specific case study. We will share some insights and recommendations concerning the GMB from this experience and will close this research report in an effort to provide specific recommendations for the improvement of the GMB process in certain similar situations.

2 OVERVIEW OF THE PRINCIPLES OF GROUP MODELING
When dealing with the GMB process, one typically refers to three key issues: (1) the choice of the group participants (including the number of participants), (2) the procedure for preparing the group sessions and (3) the facilitation of the group sessions.

2.1 The choice of the group participants

According to Vennix [7] and Andersen [8], the most important aspect of any modeling process is the selection of the right people who will participate in the model-building endeavor,

The two main issues concerning the selection of participants in a model-building process are the number of people to involve and how diverse should the group be. The recommendation with respect to these issues is to include those that have the power to act, meaning those that can implement a decision [7]. Vennix [7] also suggests the number of five participants in a GMB as the best size from his experience, but each case needs to be dealt with individually. The larger the size of the group, the more structured the sessions need to be. Andersen [8] emphasizes that it is necessary to include in the group those who have either the support for the effort that is needed, like top management, or those who will carry forward the results of the process. In summary, the suggestion is to choose different stake holders from a variety of backgrounds and culture in order for the modeling to be more efficient and beneficial [8].

2.2 The procedure of preparing the sessions

In preparing for a session, there are some considerations that should be taken into account. These include assigning different roles to group members, defining the purpose and the outcome of the session, and planning and carrying out all the logistics.

There is a general agreement in the system dynamics literature about five essential roles in the GMB process [7–9]. These five roles are: the facilitator who needs to pay attention continuously to the group processes and has to focus on the tasks of drawing out tacit knowledge and insights from the group; the modeler/reflector, who reflects on the information he/she sketches and feeds this back to the group; the process coach who usually reflects on his/her findings and provides this information to the facilitator during breaks and helps the facilitator to find ways to keep the group modeling process effective; the recorder, who writes down or sketches the important parts of the group proceedings and the gatekeeper who is the champion of the project and does all the preparatory work [9].

Several guidelines that are useful in planning the agenda are provided [7]. The first stage of the session is the introduction of all participants. The next step is to discuss the agenda. It is important to find out if there is a consensus in the group about the problem that needs to be modeled. The problem definition should be recorded and placed where everybody can see it. In case this is not the first session, reports and conclusions from previous sessions need to be provided. Clarifying what is expected from the group in this session and what outcomes are anticipated is important for participants so as to reduce anxiety. It is important to ensure that there are facilities that enable the recording of what the group is designing, and as a general rule it is advised not to write anything before ensuring that the group agrees on it. It is advisable to have the group cycle back and forth between the problem and the model. This means that there can be silences when people reflect on what has been accomplished and on how the group ought to proceed. Breaks are important to plan ahead. Finally, it is important to record preliminary and final conclusions and insights and leave the participants with a simple but clear picture of the insights, which were gained through the GMB process [7].

Others recommend planning the time so that the needs of those present are met, while time availability and the purpose of the intervention are considered [9]. There should always be room for flexibility. They believe that planning for 15 minutes blocks of time, keeps the group alert, on task, and helps to make progress. They also advise to allocate time for the members of the group so as to develop a group sense. They provide some examples of 'ice breaker' exercises, and recommend working closely with the gatekeeper to engineer the composition of small groups so that cliques are avoided. Their final recommendation is to allocate the last hour or half hour to summarize the whole day's effort in order to build a climax and to leave the conference with a feeling of accomplishment.

2.3 The facilitation of the group sessions

A group facilitator is a person who assists a group in its effort to accomplish its tasks. His concerns are about the process and the structure of the work that is being done rather than the content of the work [10]. In the case of system dynamics, the facilitator needs to have a thorough knowledge of system dynamics and extensive model-building skills in order to be able to ask the right questions during meetings [7]. Some argue that one of the most important tasks of a facilitator is to see and understand the group life (the relationship between the group members, their interdependent and interconnectedness) [10]. By understanding what is going on in the group, the facilitator is able to guide the group in a more productive way by being flexible and accommodating the needs of the group members.

The main responsibility of a facilitator is to help group members. In order to help, one needs to be patient and take enough time to learn the client's problem precisely [7]. Authenticity and integrity are of great importance to build trust between participants and facilitator. A facilitator must be neutral with respect to the content of the discussion otherwise he/she turns to be one of the group members and not an outsider, and thus, his role becomes useless. A facilitator needs to show no preference to participants in order to create a climate of trust [7].

3 THE GROUP MODELING PROCESS AT THE FIRM

A field research was conducted at the facility of a firm's subsidiary, which will be called 'the firm' in this paper. Sadia [6] described that the firm produces powdered food blends through sensitive packaging which require high level of quality standards. Concurrently, the firm also needs to maintain high level of effectiveness in order to remain profitable in the global economy where for this line of business, the competition is very intense and remaining profitable is difficult. Further details and information can be found in Sadia's work [6].

The concurrent objectives of maintaining a high level of quality standards, while at the same time minimizing production costs, have obviously created a host of problems for the firm's management. The two conflicting objectives, i.e. reducing costs and facilitating a quality culture, compelled the firm's management to seek a better way to lead in order to attain both objectives simultaneously, rather than compromising one for the other. The requirement of keeping costs low, which leads to temporary employment of some of the employees and low salaries for all of them, brings about a lot of pressure, which in turn creates problems maintaining the level of production quality and lately impacts profitability negatively.

The application of the system dynamics approach to analyze and understand the problems of the firm requires the involvement of all concerned parties within a GMB process. In order to assure that the right individuals are included in the group, it was decided to include only workers with direct connection to production. This meant workers who were included on the team were from the production, maintenance, quality, inventory and operation departments. The group consisted of nine workers including the CEO, the operation manager, the maintenance manager, the inventory manager, a food engineer, the quality manager, a quality assurance employee and two production employees. Hereinafter is an overview of the GMB process [6].

3.1 Definition of the problem in the firm

The first phase of building the conceptual model included articulating the problem of the organization as it was perceived by the participants. The articulated problem is described in Figure 1. The model includes five feedback loops. The blue arrow means that both variables

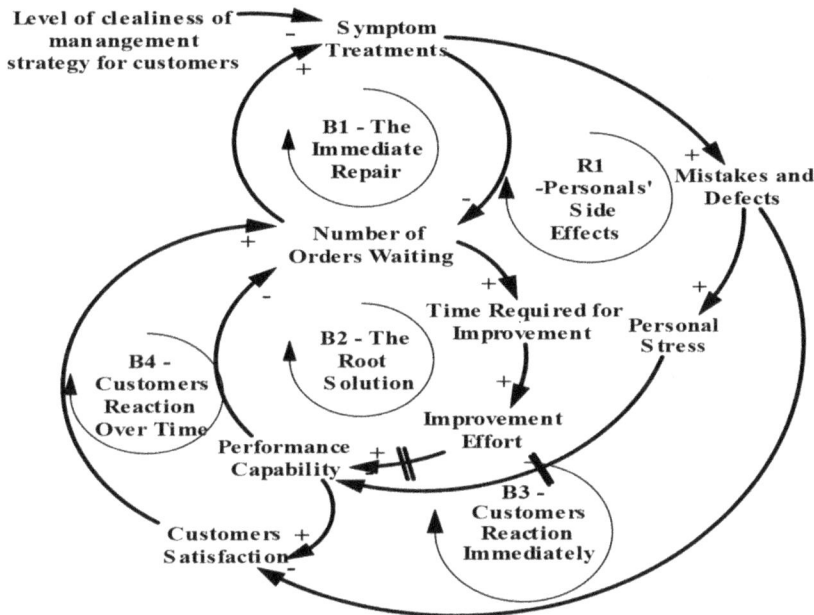

Figure 1: Conceptual model for the problem definition.

behave towards the same direction, either growing or decreasing. The red arrow means that the two variables are contradictory in their behavior. The loop direction, whether being a reinforcing loop (R) or being a balancing loop (B), can be revealed from the arrows' color (an even number of red arrows is a reinforcing feedback loop, whereas an odd number of red arrows is a balancing feedback loop). The model shows the current situation (feedback loop B1), the side effects caused by this behavior (feedback loops R1, B3, B4) and the real solution to the problem (feedback loop B2).

The 'Immediate Repair' feedback loop (B1) was considered as the main issue to deal with. The immediate response to this problem as conceived by the group participants was to push the workers to work harder in less time. This response actually solved the current problem, as described by 'The Immediate Repair' balancing feedback loop. However, such an immediate response caused some side effects.

'Side Effects for Personals' feedback loop (R1) described the side effects on personal health: the more 'Symptom Treatments' involved, the more 'Mistakes and Defects' occurred, causing 'Personal Stress' to increase which led to 'Performance Capability' to decrease and 'Number of Orders Waiting' to increase even more, causing more 'Symptom Treatments' to occur and so on, leading a reinforcing feedback loop.

'Customers Reaction Immediately' feedback loop (B3) was another side effect to 'The Immediate Repair' feedback loop: Increased 'Mistakes and Defects' decreased 'Customers Satisfaction' and therefore fewer orders were handled, decreasing the 'Numbers of Orders Waiting'. 'Customers Satisfaction' decreased also because of the decreasing of the 'Performance Capacity', as described in 'Customers Reaction Over Time' feedback loop (B4).

The root solution to the problem according to the group participants was described through the feedback loop 'The Root Solution' (B2). The accumulating 'Number of Orders Waiting' increased 'Time Required to Improvement' causing an increase to 'Improvement Effort' and after a delay – a increase of 'Performance Capacity' causing a decrease of 'Number of Order Waiting'.

From the conceptual model of the problem, one can see that when management's strategy is unclear, the workers tend to look for immediate solutions in order to solve the problem, but on the other hand, it causes side effects. Using a long-term view, the model shows that the investment in improving the process has a balancing feedback loop (B2) affecting the whole system to create more positive situations.

4 INSIGHTS REGARDING THE GMB AND POSSIBLE IMPROVEMENTS

The process of articulating the problem and building the model concurrently follows most of the literature guidelines [7, 8, 11]. In this section, the details with respect to these guidelines used will be reviewed. Further, additional elements that were included during the group modeling process will be explained, insights gained will be presented and improvements to the group modeling process will be proposed.

During the beginning stage one has to decide how many participants should be involved in the model building sessions. As mentioned before, the number of five participants in a GMB process is the best size recommended based on accumulated experience [7]; however, each case needs to be dealt with specifically. In this research, the number of participants was chosen to be nine and the sessions had to be highly structured. The group's diversity is advantageous to the model's quality but might create more tension within the group [7]. This conclusion is also reached in this research, as the diversity of the group in this research caused lots of tension and friction within the group, notably with the CEO. Therefore, the facilitator had to be very skillful at conflict resolution when such tensions arose to ensure that the process of building the model did not fall apart. In this case, the participants were chosen from all the company's levels, and as expected when different people in an organization have different interests, these differences are the sole cause of a variety in interpretation [12].

Another stage of the GMB process deals with the preparation of sessions. There were limitations when trying to use the five essential roles [7–9]. In this case, the group participants were not interested in taking on a specific role as mentioned before, therefore the facilitator had to take on some of the roles herself most of the time. In similar cases, where the participants may not be willing to undertake certain roles in the group modeling process, the five essential roles are not usable and not suitable.

The sessions were scheduled according to suggestions [7]; however, there were several important issues that are highlighted as follows:

- Since the group was larger than usually recommended (nine participants) it was deemed useful to hand out rules of conduct that explained their commitments and expected behaviors.
- A feedback form was drafted and critiqued by the group participants. Based on the critique an accepted form was used in all the sessions in order to be able to improve discussions and sessions. The feedback form included questions about the session's contents, its process and personal questions.
- Participation in a group building-model requires a high level of communication. It was important to introduce to the group some concepts on better communication skills [13].
- Since the sessions were scheduled once every three to four weeks, it was deemed necessary to provide a summary of the previous session.
- Specifically for this firm, it was found important to discuss with the group the concept of 'wants' – understanding what someone desires for him and for others that helps her/him to act towards or away from it [13]. Other kinds of problems might need different insights

and openness (e.g. the difference between feelings and thoughts), since system dynamics by its nature requires many levels of knowledge and understanding.

- In this case, we had to be very careful with the knowledge elicitation process from the group participants. Since the participants in the group were from diverse backgrounds in terms of their culture, economic situation and their work position, they could easily be led by their social condition in their thinking and opinions. Therefore, the process of knowledge elicitation through the group interlocution during the session and on the other hand, knowledge elicitation in a personal conversation brought up contradictory information and called for a continuous improvement process of knowledge elicitation through all this research.

In general, although it is preferable to plan each session in detail, it is important to be flexible during each session, to listen to the participants' intents and desires, since the participants are the facilitator's customers.

The following are recommendations for improving GMB process as the learned lessons from this experience, especially when the group members' background and their personal culture are taken into account:

- If the participants are from a diverse background, culture and education, it is recommended in such cases, to elicit knowledge and information from individuals besides the GMB session.
- The larger the number of group members, the more controlled and prepared the session should be and also less flexible. This is even more important when the level of education is diverse.
- Group modeling is also a learning process both on the personal and the team level, which requires many times raising topics that can help in self-development, improving learning abilities and raising self-awareness, i.e. feelings, wants, thoughts and better communication.
- Guiding the team work is essential and paramount in this endeavor, but no less is the personal connection and acquaintance with each of the group members and other workers, who have a huge influence on achieving data, information, thoughts and opinions in order to better understand the system's behavior.

Last but not least, the key learned issues from this research in which the existing techniques of GMB process were used, brought about some distinct insights concerning the way this process is performed. These insights ensue mainly from the dissimilarity of the culture and social background of the organization and the workers, the difference in educational level of the employees and their occupational status. The more the diversity, the more flexible techniques and facilitation approaches are needed, in order to deal with the special characteristics.

5 CONCLUSIONS

In this paper, key methodologies of the GMB process and the way they were performed in the case study were reviewed. The exploration of the differences between the theoretical frameworks of the group modeling process and its actual implementation in a case study indicate that the environment, culture and work conditions have an important impact on the way the process is followed and achieved.

Therefore, attention should be given to the dissimilarity of the culture and social background of the organization and the workers, to the difference in educational level of the employees and their occupational status, while building the conceptual model through the steps of the group modeling project. Tools that are offered here for improvement of the GMB process include establishing rules-of-conduct, learning and improving communication skills within the group, constructing the sessions and using various forms in the process and adding personal conversations with the participants. The process of knowledge elicitation has to be used by both the GMB process and personal conversations, in order to reveal underlying information about 'how things are done' in the organization.

REFERENCES

[1] Sterman, D.J., *Business Dynamics: Systems Thinking and Modeling for a Complex World*, Irwin-McGraw-Hill: New York, 2000.

[2] Ford, N.D. & Sterman, D. J., Expert knowledge elicitation to improve formal and mental models. *System Dynamics Review*, **14**(4), pp. 309–340, 1998.
https://doi.org/10.1002/(SICI)1099-1727(199824)14:4<309::AID-SDR154>3.0.CO;2-5

[3] Rouwette, E.A., Korzilius, H., Vennix, J.A.M. & Jacobs, E., Modeling as persuasion: the impact of group model building on attitudes and behavior. *System Dynamics Review*, **27**(1), pp. 1–21, 2011.

[4] Berard, C., Group model building using system dynamics: an analysis of methodological frameworks. *The Electronic Journal of Business Research Methods*, **8**(1), pp. 35–45, 2010.

[5] Rouwette, E.A., Vennix, J.A.M. & Van Mullekom, T., Group model building effectiveness: a review of assessment studies. *System Dynamics Review*, **18**(1), pp. 5–45, 2002.
https://doi.org/10.1002/sdr.229

[6] Sadia, R., A system dynamics approach linking employee health, quality culture and organizational effectiveness. Virginia: Ph.D. Dissertation, *Virginia Polytechnic Institute and State University*, 2006.

[7] Vennix, J.A.M., *Group Model Building*, John Wiley & Sons: New York and New-York, 1996.

[8] Andersen, F.D. & Richardson, P.G., Scripts for model building. *System Dynamics Review*, **13**(2), pp. 107–130, 1997.
https://doi.org/10.1002/(SICI)1099-1727(199722)13:2<107::AID-SDR120>3.0.CO;2-7

[9] Richardson, P. G. & Andersen, F. D., Teamwork in group model building. *System Dynamics Review*, **11**(2), pp. 113–137, 1995.
https://doi.org/10.1002/sdr.4260110203

[10] Phillips, D.L. & Phillips, C.M., Facilitated work groups: theory and practice. *Journal of Operational Research Society*, **44**(6), pp. 533–549, 1993.
https://doi.org/10.1057/jors.1993.96

[11] Wolstenholme, F.E., Qualitative vs. quantitative modeling: the evolving balance. *Journal of Operational Research Society*, **50**, pp. 422–428, 1999.
https://doi.org/10.1057/palgrave.jors.2600700

[12] Vennix, J.A.M., Group model building: tackling messy problems. *System Dynamics Review*, **15**(4), pp. 379–401, 1999.
https://doi.org/10.1002/(SICI)1099-1727(199924)15:4<379::AID-SDR179>3.0.CO;2-E

[13] Miller, S. & Miller, A.P., *Core Communication, Skills and Processes*, Interpersonal Communication Programs, Inc.: Littleton and Co, 1997.

RAYLEIGH–TAYLOR INSTABILITY FOR ACCELERATED METAL PLATES

ZHOU HAIBING, XIONG JUN, ZHANG SHUDAO
Institute of Applied Physics and Computational Mathematics, Beijing, China

ABSTRACT

The condensed-matter physics is changing in response to a new relevance of concepts associated with complexity. Typically, advanced properties of materials are focused on multi-scales, non-equilibrium and non-linearity. The features are also common to several other complex systems. Previous studies of the Rayleigh–Taylor instability problem for metals have been restricted to determining the cutoff wavelength in the limit of Atwood number of unity. An approximate analytical dispersion relationship is showed for elastic-plastic acceleration-driven instability growth. Here, we rigorously derive the dispersion relation for metal interface and perform a systematic investigation to compute the growth rated for the given wavelength, shear modulus and von Mises yield stress. The growth of perturbation agrees well with the experimental data for the result of numerical simulation of Rayleigh–Taylor instability of the metal driven by high-explosive detonation. It is produced that the beginning of instability can only arise via monotonically growing disturbances. A cutoff wavelength occurs for Rayleigh–Taylor instability of the metal. It is stable for the growth of perturbation. The growth of perturbation increases quickly as the perturbation wavelength increases. The most unstable wavelength and growth rate pairs are determined to cover the range of shear modulus. A mechanism is suggested to explain the wavelength selection process in recent experiments. Our model allows for including more complex physics.
Keywords: complex systems, interface instability, nonlinear dynamics, numerical simulation

1 INTRODUCTION

The Rayleigh–Taylor instability occurs when the acceleration vector points toward the heavier fluid from the lighter fluid at the initial perturbation between two fluids. The formation of spikes of denser fluid penetrates the lighter fluid and the bubbles of lighter fluid rise into the denser fluid at late times. The Rayleigh-Taylor instability can also exist in the solid medium that are accelerated by a lower-density fluid [1]. The linearized stability problem for this configuration was first treated by Lord Rayleigh. Formally identical results were later obtained by Taylor for the case of uniformly accelerated fluid layers in the presence of a gravitational field. The general problem, usually known as Rayleigh–Taylor instability, has been widely treated in the literature. In its simplest form, the Rayleigh–Taylor instability analysis treats two incompressible inviscid fluid layers of constant densities in initial equilibrium with a gravitational field. We wish to consider the instabilities, which occur when a solid body that is initially stress-free is accelerated by surface tractions. It is an important role for the material strength that the perturbation growth would be stable or unstable in the solid state.

There are many theoretical, experimental and numerical studies for the problem of Rayleigh-Taylor instability. The problems of two Newtonian fluids of different densities have been focused for most of the investigations [2]. The growth rate of classical inviscid fluid can be reduced in some conditions, such as the viscosity, the surface tension and the density gradient [1]. However, so far we are not clear understanding of the Rayleigh–Taylor instability problem for metal material. It is an important property for real media, such as strength, compressibility, viscosity and phase transitions. There are contradictory conclusions in the published papers about the consequence of the main factors, including initial wavelength, perturbation amplitude and material strength properties and loading history [3].

The first theoretical study of the Rayleigh–Taylor instability problem in metal material was performed by Miles. He thought about the infinite extent metal plate but finite thickness [2]. Robinson and Swegle reviewed the work published by other authors before 1989, and presented a method to analyse the growth of disturbances in elastic media. Some of the basic characteristic processes were explained by them [2]. The linear stability of a metal-fluid interface was elucidated by Bakhrkh for two-dimensional disturbances [3]. He derived an equation that there was a function about the growth rate and perturbation wave number. Piriz presented an analytical model approximately for the Rayleigh–Taylor instability that the solid/solid and solid/fluid interfaces were applied [5].

A set of experiments was achieved by Barnes for the Rayleigh–Taylor instability problem in metal. A thin aluminum plate was accelerated by the pressure of gaseous high-explosive detonation produces expanding across a void. The sinusoidal perturbations were machined on the surface of the aluminum plate [6]. In first shot, the perturbation growth was measured on 1100-0 Al plates of thickness 2.54mm, sinusoidal perturbations wavelength 5.08mm, and initial amplitude 100μm. In second shot, the significantly less growth was showed with half the wavelength and half the initial amplitude. In third shot, the very little growth was observed that the initial amplitude of the perturbation was halved from the first shot. Using elastic-plastic hydrodynamic code, He discussed the Rayleigh–Taylor instability of Al plates driven by high-explosive detonation [7]. The result of simulation agreed well with Barnes' experiments. It is stable for the perturbation of short wavelength. The growth of perturbation increases quickly as the perturbation wavelength increases.

Firstly, in this paper, we present a set of equations combining some effects, such as the perturbation wavelength, perturbation amplitude, material strength properties and loading history. A final approach is then given, which unites various aspects of the two previously described elastic-plastic analyses and yields a fairly simple self-consistent equation of motion. The models do provide the approximate techniques for understanding the effects of solid properties on instability response. Secondly, using elastic-plastic hydrodynamic code, we discuss the Rayleigh–Taylor instability of accelerated metal plates with the material strength properties. The results of numerical simulations are measured to both experimental data and theoretical analysis. And then the numerical simulations produce excellent agreement with experimental data. A cutoff wavelength occurs for Rayleigh–Taylor instability of the metal. We perform a systematic investigation to compute the growth rated for the given perturbation wavelength, perturbation amplitude and von Mises yield stress. Increasing yield strength retards instability growth. The regions of stability and instability separated by well-defined stability boundaries exist in the amplitude-wavelength plane.

2 ANALYTICAL MODEL

Most of the analytical models developed in the past for studying the effect of the solid properties on the Rayleigh–Taylor instability growth rate in an accelerated plate have been based on an energy balance equation [5]. The momentum equation is

$$\rho \dot{\mathbf{v}} = \nabla \cdot \mathbf{T}, \tag{1}$$

where ρ is the density, $\mathbf{v}$ is velocity and $\mathbf{T}$ is the Cauchy stress tensor. The decomposition of the Cauchy stress tensor into spherical and deviatoric components is given by

$$\mathbf{T} = -p\mathbf{I} + \mathbf{S}, \tag{2}$$

where p is the mean pressure and $\mathbf{S}$ is the deviatoric part of the stress, i.e. $\mathbf{S \cdot I} = 0$. The rate-of-strain tensor $\mathbf{D}$ is defined by

$$\mathbf{D} = \frac{1}{2}(\nabla \mathbf{v} + \nabla \mathbf{v}^T). \tag{3}$$

Incompressibility implies that $\mathbf{T:D=S:D}$. We now review various forms for the dependence of the deviatoric stress on then kinematics. For an incompressible Newtonian fluid, we have

$$\mathbf{S} = 2\mu \mathbf{D}, \tag{4}$$

where μ is the viscosity. For an elastic incompressible material, we have

$$\dot{\mathbf{S}} = 2G\mathbf{D}, \tag{5}$$

where G is the elastic shear modulus. For an elastic-plastic incompressible material, we shall assume a Prandtl-Reuss flow rule and von Mises yield stress criterion of the form

$$\dot{\mathbf{S}} + \Lambda\mathbf{S} = 2G\mathbf{D} \tag{6}$$

$$\Lambda = G\dot{W} / s_1^2, \text{ provided } J_2 = 2\,s_1^2 \text{ and } \dot{W} > 0 \tag{7}$$

$$\Lambda = 0, \text{ provided } J_2 < 2\,s_1^2 \text{ or } \dot{W} < 0 , \tag{8}$$

where $\dot{W} = \mathbf{S \cdot D}$ and $J_2 = \mathbf{S:S}$. We emphasize here that in the context of the incompressible analysis of this paper, an elastic response should be interpreted as an elastic shear response, not, of course, as an elastic volumetric response.

The energy balance equation is obtained by multiplying eqn. (1) by $\mathbf{v}$ and then integrating over the volume.

$$\int \rho \dot{\mathbf{v}} \cdot \mathbf{v} d\Omega = \int \nabla \cdot \mathbf{T} \cdot \mathbf{v} d\Omega. \tag{9}$$

Use of the divergence theorem and the symmetry of the stress tensor leads to eqn. (10).

$$-\oint p\mathbf{n} \cdot \mathbf{v} dS - \int \rho \dot{\mathbf{v}} \cdot \mathbf{v} d\Omega = \int \mathbf{S:D} d\Omega \tag{10}$$

The surface of plates has a sinusoidal perturbation.

$$y = q \cos kx, \tag{11}$$

where q is the amplitude of perturbation, k is the wave number of perturbation. The motion is assumed to be periodic in x of wavelength λ and wave number $k=2\pi/\lambda$ with an imposed zero stress boundary condition on the upper surface and a various pressure on the lower surface, $p(t)$. We assume that the kinematics of the flow follows the plane incompressible irrotational motion. The velocity field is given by

$$\mathbf{v} = \dot{q}e^{-ky}\begin{pmatrix} \sin kx \\ \cos kx \end{pmatrix} \tag{12}$$

The volume V is taken over one wave length. Evaluating each term on the eqn. (10) yields

$$-\oint p\mathbf{n} \cdot \mathbf{v} dS = \pi p(t)q\dot{q} \tag{13}$$

$$-\oint \rho \dot{\mathbf{v}} \cdot \mathbf{v} d\Omega = -\frac{\rho \lambda \dot{q} \ddot{q}}{2k}\left(1 - e^{-2kh}\right) \tag{14}$$

where h is the thickness of plates.

The analysis follows as usual by evaluating right side of eqn. (10) with S given by eqn. (4, 5, 6). in the elastic case, eqn. (5) holds. eqn. (10) simplifies to

$$\ddot{q} + \frac{k^2}{\rho}\left(4G - \frac{p(t)}{1 - e^{-2kh}}\right)q = 4G\frac{k^2}{\rho}q_0. \tag{15}$$

When the pressure is constant, $p=\rho ah$, the cutoff perturbation wave number is

$$k_c = \frac{\rho a}{8G}. \tag{16}$$

The result is similar to the one obtained originally [4, 5].

The stability response equations derived from incompressible analyses are capable of capturing the effects which are due to wave propagation and also predict the growth due to time-dependent effects. The basic point of agreement is that both the material strength and loading history play an important role in determining the location of the stability-instability boundary in an initial amplitude-wavelength stability diagram.

3 NUMERICAL TREATMENTS

In the Barnes' experiments, thin plates with a perturbation machined in the driven surface were accelerated by gaseous detonation products from a plane wave generator. A gap separated the plane wave generator and the perturbed surface of the plate, so that a shock was not directly transmitted into the plate, and smooth acceleration was obtained. The configuration is shown in Figure 1. There is a sinusoidal perturbation of a given wavelength (λ) and amplitude (A) on the surface of the Al plate (thickness, h). The perturbation surface is driven by the pressure of high-explosive detonation produces, while the opposite surface is flat [6, 8].

3.1 Numerical Method

The response of aluminum plates are calculated by numerical code, CHAP. It is a general purpose elastic-plastic hydrodynamic code. The main discrete spatial form is based on the

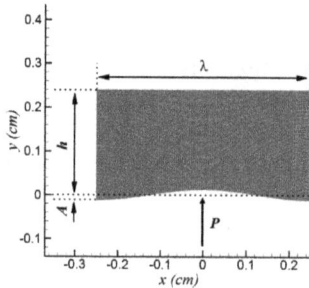

Figure 1: Configuration used in the plate stability study, showing the plate thickness h, the perturbation amplitude A, the perturbation wavelength λ, and the application of the driving pressure P [9]

compatible Lagrangian numerical method. The predictor-corrector method, an explicit time integration, is applied to the discrete time form. The spurious motions are resisted by the subzonal pressure. The hourglass distortion is controlled effectively. The tensor artificial viscosity automatically captures shock wave discontinuities. The material strength and high-explosive reaction are achieved currently. The many simulation have been presented. CHAP can be applied to numerical studies of complex engineering issue [7, 10].

The Euler method is used for the classical Rayleigh–Taylor instability problem because there are velocity shear and turbulent mixing. But, in this paper we focus on the Rayleigh–Taylor instability of metals. The pressure varies little and the deformation of metal is little. The advantage is the more resolution of interface for Lagrangian numerical method [7, 10].

3.2 Calculated Approach

In previous simulations of these experiments, the ram pressure induced by the pile-up of the high-explosive detonation products was treated as a time-dependent pressure drive uniformly distributed over the plate surface [1]. We took the same approach, using the pressure history shown in Figure 2.

The unit of material model is centimeter, gram and microsecond in what follows. The density of aluminum plates is 2.78.

The Mie-Gruneisen EOS is used:

$$p = \frac{\rho_0 c^2 \mu \left[1 + \left(1 - \frac{\gamma_0}{2} \right) \mu - \frac{a}{2} \mu^2 \right]}{\left[1 - (S_1 - 1) \mu - S_2 \frac{\mu^2}{\mu+1} - S_3 \frac{\mu^3}{(\mu+1)^2} \right]} + (\gamma_0 + a\mu) E, \tag{1}$$

where μ is the relative compression, $\mu=(1-V)/V$, V is the relative volume (current volume per initial volume), E is the internal energy per initial volume, ρ_0 is the initial density, the coefficients $c=0.5328$, $S_1=1.338$, $S_2=0.0$, $S_3=0.0$, $\gamma_0=2.0$, $a=0.0$.

The Steinberg constitutive model is used:

$$G = G_0 \left[1 + bpV^{1/3} - h \left(\frac{E - E_c}{3R'} - 300 \right) \right] e^{-\frac{fE}{E_m - E}} \tag{2}$$

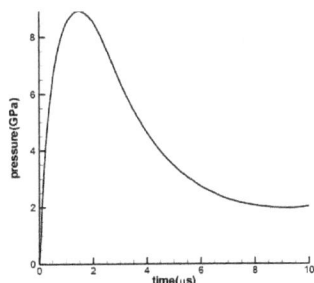

Figure 2: Applied pressure profile used to simulate the metal plates [7]

$$\sigma_y = \sigma_0' \left[1 + b'pV^{1/3} - h\left(\frac{E - E_c}{3R'} - 300 \right) \right] e^{-\frac{fE}{E_m - E}} \tag{3}$$

$$\sigma_0' = \sigma_0 \left[1 + \beta \left(\gamma_i + \varepsilon_p \right) \right]^n , \tag{4}$$

where the shear modulus G_0=0.276, the yield strength σ_0=0.002, the hardening coefficient β=50, the hardening exponent n=0.05, the shear modulus hardening coefficient b=8.0, the yield strength hardening coefficient b'=8.0, the initial plastic strain γ_i=0.0, the temperature softening coefficient h=0.0, the maximum yield strength Y_{max}=0.0068; the melting temperature T=1220.

4 RESULTS AND DISCUSSION

4.1 Comparison with Experimental Results

In order to establish the validity of a numerical study, the calculations should be compared to experimental result. The experimental data [6] were obtained for two initial configurations. The first had an applied perturbation whose wavelength was twice the 2.54mm plate thickness, and whose amplitude (half the total peak-to-trough surface variation) was 0.102mm. The other configuration had half the amplitude, one with one-half that wavelength. There are two models. Model one is h=2.54mm, λ=5.08mm, A=0.102mm; Model two is h=2.54mm, λ=2.54mm, A=0.050mm [9].

The perturbation growth factor histories are focused. The perturbation growth factor is defined that the current amplitude is divided by initial perturbation amplitude. In the other word, the perturbation growth factor shows that the current amplitude is the multiple of initial perturbation amplitude.

Figure 3 shows the comparison of experimental peak-to-trough surface variation data with calculated results for aluminum plates driven by the pressure history of Figure 2. In Figure 3, the black squares are results of experiment by Barnes [6] and the solid curves are the results of simulation by CHAP. The simulated growth factor corresponds to the measured one. As can be seen, excellent agreement is obtained using an elastic-plastic model for the aluminum.

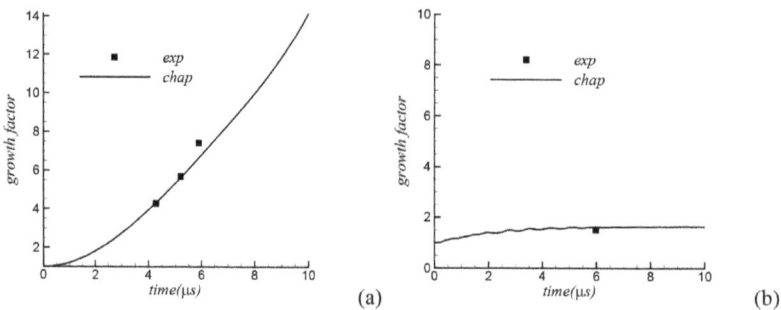

Figure 3: Comparison of experimental and calculated results for perturbation growth in accelerated aluminum plates [7] (a) model one, (b) model two.

Thus, the calculations have been correlated with experimental stability data, and the excellent agreement provides a calibration point for the calculated results. A cutoff wavelength occurs for Rayleigh–Taylor instability of the metal. It is stable for the growth of perturbation when the perturbation wavelength is less than the critical wavelength [9].

4.2 Calculated Perturbation Growth

Once calculations have been done for the configurations for which experimental data exist, and good agreement between calculation and experiment has provided confidence in the calculations, the numerical study can be expanded in order to determine the qualitative property of the phenomenon of instability growth in accelerated plates and to quantity the dependence on various parameters.

Figure 4 demonstrates the effect of varying the wavelength of the perturbation. For a given yield strength, a wavelength exists for which the surface variation grows most rapidly. A cutoff wavelength occurs for Rayleigh–Taylor instability of the metal. It is stable for short wavelength. The growth of perturbation increases rapidly as the perturbation wavelength increases.

Figure 5 shows the effect of varying the yield strength. As can be seen from the figure, increasing the yield strength retards the growth of perturbation. The higher yield strength

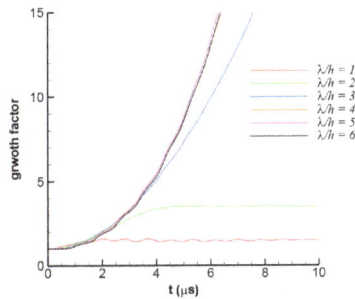

Figure 4: Effect of wavelength variations on surface. initial amplitude A/λ=0.5%. the maximum yield strength Y_{max}=0.002.

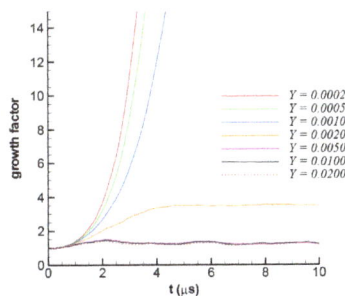

Figure 5: Effect of yield strength variations on surface. the perturbation wavelength λ/h=2, initial amplitude A/λ=0.5%.

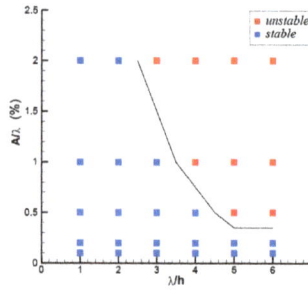

Figure 6: Results of several numerical calculations, demonstrating the mapping of the stability boundary (solid line) in the amplitude-wavelength plane. the maximum yield strength Y_{max}=0.002.

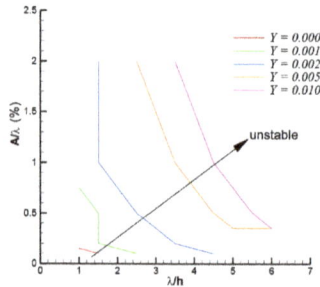

Figure 7: Stability boundaries at various yield strengths in the amplitude-wavelength plane.

calculations become stable in that the perturbation ceases to grow after the pressure becomes constant.

In reality neither amplitude nor wavelength alone controls stability. Instead, a boundary, which is both amplitude and wavelength dependent, separates regions of stability and instability, as shown in Figure 6. As can be seen, the amplitude that remains stable decreases as the wavelength increases. Further wavelength increases show the existence of a cutoff wavelength.

Figure 7 shows stability boundaries of varying the yield strength. increasing yield strength increases the stable amplitude. The cutoff wavelength appears to be dependent of yield strength.

5 CONCLUSION

An attractive approach without conscious reasoning has been presented to the Rayleigh–Taylor instability with the purpose of developing a method for dealing with complex conditions that exist in accelerated metal media. The asymptotic response of the perturbation growth is reproduced by the present model. The model shows that the growth of a perturbation depends on its amplitude, wavelength, driving pressure history and material property. The model gives the impression to be very appropriate for dealing with more complex conditions.

Using elastic-plastic hydrodynamic code (CHAP), We discuss the Rayleigh–Taylor instability of Al plates driven by high-explosive detonation. The numerical method produces excellent agreement with experimental data.

It is important that the elastic-plastic role must be supposed for the numerical result of Rayleigh–Taylor instability of the metal driven by high-explosive detonation. A cutoff wavelength occurs for Rayleigh–Taylor instability of the metal. It is stable for the growth of perturbation. The growth of perturbation increases quickly as the perturbation wavelength increases. Increasing yield strength retards instability growth. The regions of stability and instability separated by well-defined stability boundaries exist in the amplitude-wavelength plane.

REFERENCES
[1] Colvin, J.D., Legrand, M., Remington, B.A., Schurtz, G. & Weber, S.V., A model for instability growth in accelerated solid metals. *Journal of Applied Physics*, **93**(9), pp. 5287–5301, 2003. doi: 10.1063/1.1565188

[2] Terones, G., Fastest growing linear Rayleigh-Taylor modes at solid/fluid and solid/solid interfaces. *Physical Review E*, **71**(3), 036306. doi:10.1103/physreve.71.036306

[3] Bakhrakh, S.M., Drennov, O.B., Kovalev, N.P., Lebedev, A.I., Meshkov, E.E., Mikhailov, A.L., Nevmerzhitsky, N.V., Nizovtsev, P.N., Rayevsky, A.A., Simonov, G.P., Solovyev, V.P. & Zhidov, I.G., *Hydrodynamic instability in strong media*. Technical Report UCRL-CR-126710, Lawrence Livermore National Laboratory, 1997.

[4] Robinson, A.C. & Swegle, J.W., Acceleration instability in elastic-plastic solids. II. Analytical techniques. *Journal of Applied Physics*, **66**(7), pp. 2859–2872, 1989. doi:10.1063/1.344191

[5] Piriz, A.R., Lopez Cela, J.J. & Cortazar, O.D., Rayleigh-Taylor instability in elastic solids. *Physical Review E*, **72**(5), 056313. doi:10.1103/physreve.72.056313

[6] Barnes, J.F., Blewett, P.J., McQueen, R.G., Meyer, K.A. & Venable, D., Taylor instability in solids. *Journal of Applied Physics*, **45**, pp. 727–732, 1974.

[7] Changjiang, H., Haibing, Z. & Yihong, H., A numerical study on Rayleigh-Taylor instability of aluminum plates driven by detonation. *Science China: Physics, Mechanics & Astronomy*, **53**(2), pp. 195–198, 2010. doi:10.1007/s11433-009-0261-4

[8] Swegle, J.W. & Robinson, A.C. Acceleration instability in elastic-plastic solids. I. Numerical simulations of plate acceleration. *Journal of Applied Physics*, **66**(7), pp. 2838–2858, 1989. doi:10.1063/1.344190

[9] Haibing, Z., Changjiang, H., Shudao, Z. & Yihong, H. The numerical studies on Rayleigh-Taylor instability of aluminum plates driven by detonation. *Proceedings of 12th International Workshop on the Physics of Compressible Turbulent Mixing*, Russia, 2010, available at http://iwpctm12.imamod.ru

[10] Shudao, Z., Haibing, Z. & Wentao, L., *Investigation and application of compatible lagrangian hydrodynamics numerical method*. GF-A0091252, Institute of Applied Physics and Computational Mathematics, 2005.

A MODIFIED HLLC METHOD AND ITS APPLICATION IN BEAR–NUNZIATO MODEL

ZHENG LI SHUANG-HU WANG
Institute of Applied Physics and Computational Mathematics, Beijing 100094, China

ABSTRACT

Wave-speed is very important for HLLC method. If the wave-speed is too small, the wave characters cannot catch; if it is too large, there is more viscosity in the calculation. There are a lot of methods to determine the wave-speed, but every method has its applicability. The prescript present a modified HLLC method without estimating wave-speed, and it is applied to simulate the Baer–Nunziato (BN) model. It is compared with three classic wave-speed estimating method in the prescript. Numerical results validate the current method for simulation of several typical BN model initial problems.

Keywords: Baer–Nunziato model; compressible; HLLC method; multiphase flow; wave-speed

1 INTRODUCTION

Computation of compressible multiphase flow problems are important in science and technology domains. A popular approach to describe two-phase flows is so-called averaged mixture model. In the two-phase model, two phases treat as continuous media, and they are uniform mixed. Baer–Nunziato model (BN model) of Bear and Nunziato [1] is one of the multiphase flow models. The governing equations consist of mass equation, momentum equation and energy equation of each phase. The BN model is a compressible and non-conservative two-phase model. Non-conservative terms describe interaction of two phases in mass, momentum and energy. The governing equations cannot write in conservation form, and that is difficult in numerical simulation.

In recent years, numerical method of BN model has attracted much attention. Embid and Baer [2] studied the mathematical character of BN model and got the eigenvalues and eigenvectors of the equations. Andrianov and Warnecke [3] construct exact solution by property of solid phase contact discontinuity. And they found that the solution maybe nonunique. Schwendeman *et al.* [4] determine the wave structure of Riemann problem by thin-layer theory beside the solid concact, which can get intermediate states by iterative method. And they construct a godunov method and an adaptive Riemann solver for computational efficiency. Deledicque and Papalexandris [5] present a new numerical procedure for solving exactly the Riemann problem by four principal configurations. Tokareva and Toro [6] construct an HLLC approximate Riemann solver for BN model.

In this paper, we analyse the method again, and put forward a modified method, which need not estimate the wave-speed. The intermediate states can be written as a function of wave-speed. The wave-speed can be derived from solving the thin-layer equations beside the solid contact by iteration. Then the intermediate states can write out directly.

The rest of this paper is organized as follows. In Section 2, we recall the governing equations and describe the features of the Riemann problem. In Section 3, we present the modified HLLC Riemann solver in details. In Section 4, we systematically assess the current method. Numerical simulation of some typical BN model initial problems are presented. we compare and analyse the results with three classical wave-speed method. Conclusion is drawn in Section 5.

2 BEAR–NUNZIATO MODEL AND ITS RIEMANN SOLUTION

2.1 Bear–Nunziato model

Consider Baer–Nunziato nonconservative multiphase model in one dimension

$$\partial_t W + \partial_x G(W) + H(W)\partial_x \bar{\alpha} = 0, \tag{1}$$

where W, G and H is

$$W = \begin{bmatrix} \bar{\alpha} \\ \overline{\alpha\rho} \\ \overline{\alpha\rho u} \\ \overline{\alpha\rho \bar{E}} \\ \alpha\rho \\ \alpha\rho u \\ \alpha\rho E \end{bmatrix}, G(W) = \begin{bmatrix} 0 \\ \overline{\alpha\rho u} \\ \bar{\alpha}(\overline{\rho u}^2 + \bar{p}) \\ \overline{\alpha u}(\bar{\rho}\bar{E} + \bar{p}) \\ \alpha\rho u \\ \alpha(\rho u^2 + p) \\ \alpha u(\rho E + p) \end{bmatrix}, H(W) = \begin{bmatrix} \bar{u} \\ 0 \\ -p \\ -p\bar{u} \\ 0 \\ p \\ p\bar{u} \end{bmatrix},$$

where α and $\bar{\alpha}$ are volume fraction of each phase, with $0 < \alpha < 1$, $0 < \bar{\alpha} < 1$ and $\alpha + \bar{\alpha} = 1$. $\bar{\rho}$, $\bar{u}$, $\bar{p}$ are density, velocity and pressure of solid phase, and its total energy $\bar{E} = \bar{e} + 1/2\bar{u}^2$, $\bar{e}$ is internal energies of solid phase. ρ, u, p are density, velocity and pressure of gas phase and its total energy $E = e + 1/2u^2$, e is internal energies of gas phase.

The above system is closed with the equation of state (EOS) for each phase. We assume an ideal EOS for the gas phase and a stiffened EOS for the solid phase:

$$p = (\gamma - 1)\rho e, \tag{2}$$

$$\bar{p} = (\bar{\gamma} - 1)\bar{\rho}\bar{e} - \bar{\gamma}\bar{p}_\infty, \tag{3}$$

where γ is specific heat ratio of gas phase. $\bar{\gamma}$ is specific heat ratio of solid phase and $\bar{p}_\infty$ is a constant. The sound speed of each phase can be written as $c = \sqrt{\gamma p / \rho}$ and $\bar{c} = \sqrt{\bar{\gamma}(\bar{p} + \bar{p}_\infty)/\bar{\rho}}$.

2.2 Riemann problem

The general structure of the solution of BN model Riemann problem in one dimension consists of shocks, rarefactions, and contacts. The wave structure of BN model is six, so it is hard to solve. Consider $(u - \bar{u})^2 < c^2$, Figure 1 shows the solution structure of Riemann problem of BN model. The volume fraction only discontinues at contact of solid gas. It shows that the

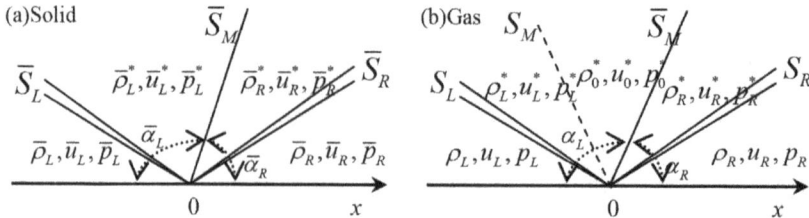

Figure 1: Solusion structure of Rieamann problem of BN model

gas phase and solid phase are independent each other except beside the contact of the solid phase. And each phase satisfies the Euler equations. There are two intermediate states *L and *R beside the contact of solid phase. The density and pressure of each intermediate state are different. But the velocity is same, $\bar{u}_L^* = \bar{u}_R^*$. The intermediate states of gas phase are separated into three parts, *L *0 and *R, by contact of gas phase and solid phase. In *0, they satisfy

$$\rho_0^* = \rho_L^* \left(p_R^* / p_L^* \right)^{1/\gamma}, u_0^* = u_R^*, p_0^* = p_R^*.$$

3 HLLC METHOD

The character of our method is that the intermediate states can be written as function of wave-speed. Then we can get the wave-speed from solving the thin-layer equations by iteration. When we get the wave-speed, the intermediate states can be written out directly. Assume wave-speed of solid phase and gas phase are $\bar{S}_L, \bar{S}_M, \bar{S}_R$ and S_L, S_M, S_R.

3.1 Intermediate states

Conservative states of each phase are $\bar{U} = [\overline{\alpha\rho}, \overline{\alpha\rho u}, \overline{\alpha\rho E}]^T$ and $U = [\alpha\rho, \alpha\rho u, \alpha\rho E]^T$; and their corresponding fluxes are $\bar{F} = [\overline{\alpha\rho u}, \bar{\alpha}(\overline{\rho u^2} + \bar{p}), \overline{\alpha u}(\overline{\rho E} + \bar{p})]^T$ and $F = [\alpha\rho u, \alpha(\rho u^2 + p), \alpha u(\rho E + p)]^T$. Then the intermediate states satisfy

$$
\begin{aligned}
\bar{F}_L^* &= \bar{F}_L + \bar{S}_L(\bar{U}_L^* - \bar{U}_L) \\
\bar{F}_R^* &= \bar{F}_R + \bar{S}_R(\bar{U}_R^* - \bar{U}_R) \\
F_L^* &= F_L + S_L(U_L^* - U_L) \\
F_R^* &= F_R + S_R(U_R^* - U_R)
\end{aligned}
\tag{4}
$$

We make the solid intermediate states of *L as an example, it satisfy

$$
\begin{cases}
\bar{\alpha}_L^* \bar{\rho}_L^* \bar{u}_L^* = \bar{\alpha}_L \bar{\rho}_L \bar{u}_L + \bar{S}_L(\bar{\alpha}_L^* \bar{\rho}_L^* - \bar{\alpha}_L \bar{\rho}_L) & (5.1) \\
\bar{\alpha}_L^* (\bar{\rho}_L^* \bar{u}_L^{*2} + \bar{p}_L^*) = \bar{\alpha}_L (\bar{\rho}_L \bar{u}_L^2 + \bar{p}_L) + \bar{S}_L(\bar{\alpha}_L^* \bar{\rho}_L^* \bar{u}_L^* - \bar{\alpha}_L \bar{\rho}_L \bar{u}_L) & (5.2) \\
\bar{\alpha}_L^* \bar{u}_L^* (\bar{\rho}_L^* \bar{E}_L^* + \bar{p}_L^*) = \bar{\alpha}_L \bar{u}_L (\bar{\rho}_L \bar{E}_L + \bar{p}_L) + \bar{S}_L(\bar{\alpha}_L^* \bar{\rho}_L^* \bar{E}_L^* - \bar{\alpha}_L \bar{\rho}_L \bar{E}_L) & (5.3)
\end{cases}
$$

From (5.1) and (5.2), we can get

$$\bar{\rho}_L^* = \bar{\rho}_L \frac{\bar{u}_L - \bar{S}_L}{\bar{u}_L^* - \bar{S}_L} \tag{6}$$

$$\bar{p}_L^* = \bar{p}_L + \bar{\rho}_L(\bar{S}_L - \bar{u}_L)(\bar{u}_L^* - \bar{u}_L). \tag{7}$$

With solid energy express $\bar{E}_L^* = \dfrac{\bar{p}_L^* + \bar{\gamma}\bar{p}_\infty}{(\bar{\gamma}-1)\bar{\rho}_L^*} + \dfrac{1}{2}\bar{u}_L^{*2}$ and $\bar{E}_L = \dfrac{\bar{p}_L + \bar{\gamma}\bar{p}_\infty}{(\bar{\gamma}-1)\bar{\rho}_L} + \dfrac{1}{2}\bar{u}_L^2$, and sound speed $\bar{c} = \sqrt{\bar{\gamma}(\bar{p} + \bar{p}_\infty)/\bar{\rho}}$, we can get

$$(\bar{u}_L^* - \bar{u}_L)\left[\frac{\bar{c}_L^2}{\bar{\gamma}-1} + \frac{1}{2}(\bar{u}_L - \bar{S}_L)(\bar{u}_L^* + \bar{u}_L) - \frac{(\bar{u}_L - \bar{S}_L)}{\bar{\gamma}-1}(\bar{\gamma}\bar{u}_L^* - \bar{S}_L) \right] = 0. \tag{8}$$

The results of equation (8) are $\bar{u}_L^* = \bar{u}_L$ and $\bar{u}_L^* = \bar{u}_L + \dfrac{2}{\bar{\gamma}+1}(\bar{S}_L - \bar{u}_L - \dfrac{\bar{c}_L^2}{\bar{S}_L - \bar{u}_L})$. If $\bar{u}_L^* = \bar{u}_L$, then $\bar{p}_L^* = \bar{p}_L$, $\bar{\rho}_L^* = \bar{\rho}_L$. It degenerates to const, so it should reject. So $\bar{u}_L^*$ can be written as function (9) of $\bar{S}_L$.

$$\bar{u}_L^*(\bar{S}_L) = \bar{u}_L + \frac{2}{\bar{\gamma}+1}(\bar{S}_L - \bar{u}_L - \frac{\bar{c}_L^2}{\bar{S}_L - \bar{u}_L}). \tag{9}$$

We put (9) into (6) and (7), $\bar{\rho}_L^*$ and $\bar{p}_L^*$ can also be written as function of $\bar{S}_L$. The solid physical variables of *R can be written as function of $\bar{S}_R$. The intermediate states of solid can written as unified form of (10), where k=L, R.

$$\bar{\rho}_K^* = \bar{\rho}_K \frac{(\bar{\gamma}+1)(\bar{u}_K - \bar{S}_K)^2}{(\bar{\gamma}-1)(\bar{u}_K - \bar{S}_K)^2 + 2\bar{c}_K^2}$$

$$\bar{u}_K^* = \bar{u}_K + \frac{2}{\bar{\gamma}+1}(\bar{S}_K - \bar{u}_K - \frac{\bar{c}_K^2}{\bar{S}_K - \bar{u}_K}). \tag{10}$$

$$\bar{p}_K^* = \bar{p}_K + \frac{2\bar{\rho}_K}{\bar{\gamma}+1}[(\bar{S}_K - \bar{u}_K)^2 - \bar{c}_K^2]$$

Similarly, gas intermediate states of *L and *R can be written as function of S_K (k=L,R).

As show in Figure 1(b),where the solid contact is at the right of the gas contact, values of *0 are

$$\begin{aligned} \alpha_0 &= \alpha_L \\ \rho_0^* &= \rho_R^*(p_L^*/p_R^*)^{1/\gamma} \\ u_0^* &= u_L^* \\ p_0^* &= p_L^* \end{aligned}. \tag{11}$$

If solid contact is at the left of the gas contact, values of *0 are (12).

$$\begin{aligned} \alpha_0 &= \alpha_R \\ \rho_0^* &= \rho_L^*(p_R^*/p_L^*)^{1/\gamma} \\ u_0^* &= u_R^* \\ p_0^* &= p_R^* \end{aligned}. \tag{12}$$

3.2 Thin-layer theory

Because the volume fraction only discontinues at constant of solid gas, so the intermediate physical states of each phase are coupled near the solid contact. In 2006, Schwendeman [4] put forward thin-layer theory to analysis the physical states situations near the solid contact. We only show the case of $(u - \bar{u})^2 < c^2$, and there are two situations. When $u < \bar{u}$, as shown in Figure 1, they satisfy (13).

$$\begin{cases} \bar{u}_{*R} - \bar{u}_{*L} = 0 \\ \alpha_R(u_{*R} - \bar{u}_{*R}) - \alpha_L(p_{*L}/p_{*R})^{1/\gamma}(u_{*L} - \bar{u}_{*L}) = 0 \\ \bar{\alpha}_R \bar{p}_{*R} + \alpha_R p_{*R} - \bar{\alpha}_L \bar{p}_{*L} - \alpha_L p_{*L} + \alpha_R \rho_{*R}(u_{*R} - \bar{u}_{*R})(u_{*R} - u_{*L}) = 0 \\ \dfrac{\gamma}{\gamma-1}[\dfrac{p_{*R}}{\rho_{*R}} - \dfrac{p_{*L}}{\rho_{*R}}(\dfrac{p_{*R}}{p_{*L}})^{1/\gamma}] + [(u_{*R} - \bar{u}_{*R})^2 - (u_{*L} - \bar{u}_{*L})^2]/2 = 0 \end{cases} \quad (13)$$

If $u > \bar{u}$, they satisfy (14):

$$\begin{cases} \bar{u}_{*R} - \bar{u}_{*L} = 0 \\ \alpha_R(p_{*R}/p_{*L})^{1/\gamma}(u_{*R} - \bar{u}_{*R}) - \alpha_L(u_{*L} - \bar{u}_{*L}) = 0 \\ \bar{\alpha}_R \bar{p}_{*R} + \alpha_R p_{*R} - \bar{\alpha}_L \bar{p}_{*L} - \alpha_L p_{*L} + \alpha_L \rho_{*L}(u_{*L} - \bar{u}_{*L})(u_{*R} - u_{*L}) = 0 \\ \dfrac{\gamma}{\gamma-1}[\dfrac{p_{*R}}{\rho_{*L}}(\dfrac{p_{*L}}{p_{*R}})^{1/\gamma} - \dfrac{p_{*L}}{\rho_{*L}}] + [(u_{*R} - \bar{u}_{*R})^2 - (u_{*L} - \bar{u}_{*L})^2]/2 = 0 \end{cases} \quad (14)$$

The intermediate states of thin-layer theory can be written as function of wave-speed $S_K(k=L,R)$, so we can get the wave-speed from solving the thin-layer equations by iteration. The initial wave-speed of most problems can choose the max–min eigenvalues (15).

$$\bar{S}_L = \bar{u}_L - \bar{c}_L, \bar{S}_R = \bar{u}_R + \bar{c}_R, S_L = u_L - c_L, S_R = u_R + c_R. \quad (15)$$

But for some problems, the initial wave-speed need determined by average pressure of Riemann problem of each phase. We can take $\bar{p}^*$ and p^* for average pressure of Riemann problem of each phase. Wave-speed can be thought function of average pressure from (10), so initial data can be written as (16) and (17).

$$\bar{S}_L = \bar{u}_L - \bar{c}_L \sqrt{\dfrac{\bar{\gamma}+1}{2\bar{\gamma}}\dfrac{\bar{p}^*+\bar{p}_\infty}{\bar{p}_L+\bar{p}_\infty} + \dfrac{\bar{\gamma}-1}{2\bar{\gamma}}}, \bar{S}_R = \bar{u}_R + \bar{c}_R \sqrt{\dfrac{\bar{\gamma}+1}{2\bar{\gamma}}\dfrac{\bar{p}^*+\bar{p}_\infty}{\bar{p}_R+\bar{p}_\infty} + \dfrac{\bar{\gamma}-1}{2\bar{\gamma}}} \quad (16)$$

$$S_L = u_L - c_L \sqrt{\dfrac{\gamma+1}{2\gamma}\dfrac{p^*}{p_L} + \dfrac{\gamma-1}{2\gamma}}, S_R = u_R + c_R \sqrt{\dfrac{\gamma+1}{2\gamma}\dfrac{p^*}{p_R} + \dfrac{\gamma-1}{2\gamma}} \quad (17)$$

3.3 Numerical Flux

As show in Figure 1(a), Solid phase is divided into four const regions. We take $\bar{S}_M = \bar{u}_L^* = \bar{u}_R^*$, its corresponding flux can be written as (18).

$$\bar{F}(\bar{U}_L, \bar{U}_R) = \begin{cases} \bar{F}_L = \bar{F}(\bar{U}_L), & 0 \le \bar{S}_L \\ \bar{F}_L^* = \bar{F}(\bar{U}_L^*), & \bar{S}_L \le 0 \le \bar{S}_M \\ \bar{F}_R^* = \bar{F}(\bar{U}_R^*), & \bar{S}_M \le 0 \le \bar{S}_R \\ \bar{F}_R = \bar{F}(\bar{U}_R), & \bar{S}_R \le 0 \end{cases} . \quad (18)$$

Gas phase can be divided into five const regions, and its flux can be written as (19).

$$F(U_L,U_R) = \begin{cases} F_L = F(U_L), & 0 \leq S_L \\ F_L^* = F(U_L^*), & S_L \leq 0 \leq \min(S_M,\bar{S}_M) \\ F_0^* = F(U_0^*), \min(S_M,\bar{S}_M) \leq 0 \leq \max(S_M,\bar{S}_M). \\ F_R^* = F(U_R^*), \max(S_M,\bar{S}_M) \leq 0 \leq S_R \\ F_R = F(U_R), & S_R \leq 0 \end{cases} \tag{19}$$

3.4 Finite volume

Assume a finite-volume method with volumes $[x_{i-1/2},x_{i+1/2}]*[t^n,t^{n+1}]$, signed Ω_i^n, cell spacing $\Delta x_i = x_{i+1/2} - x_{i-1/2}$, cell center x_i, and time step $\Delta t^n = t^{n+1} - t^n$. We define cell average $W_i^n = \int_{x_{i-1/2}}^{x_{i+1/2}} W(x,t^n)dx \Big/ \Delta x_i$ and integrate equation (1) over the control volume Ω_i^n.

$$W_i^{n+1} = W_i^n - \frac{\Delta t^n}{\Delta x_i}(G_{i+1/2}^n - G_{i-1/2}^n) - \frac{1}{\Delta x_i}\iint_{\Omega_i^n} H(W)\partial_x \bar{\alpha}\,dxdt \tag{20}$$

Where, $G_{i+1/2}^n = \left(0, \bar{F}(\bar{U}_i,\bar{U}_{i+1}), F(U_i,U_{i+1})\right)^T$, $\bar{F}$ and F are (18) and (19). We use the method of Schwendeman [4] and Toro [6] to approximate $\iint_{\Omega_i^n} H(W)\partial_x\bar{\alpha}\,dxdt \Big/ \Delta x_i$.

$$\tilde{H} = \begin{pmatrix} \bar{S}_M(\bar{\alpha}_R - \bar{\alpha}_L) \\ 0 \\ \bar{\alpha}_L\bar{p}_L^* - \bar{\alpha}_R\bar{p}_R^* \\ \bar{S}_M(\bar{\alpha}_L\bar{p}_L^* - \bar{\alpha}_R\bar{p}_R^*) \\ 0 \\ \bar{\alpha}_R\bar{p}_R^* - \bar{\alpha}_L\bar{p}_L^* \\ \bar{S}_M(\bar{\alpha}_R\bar{p}_R^* - \bar{\alpha}_L\bar{p}_L^*) \end{pmatrix} \tag{21}$$

(20) simplify as

$$W_i^{n+1} = W_i^n - \frac{\Delta t^n}{\Delta x_i}(L_{i+1/2}^- - L_{i-1/2}^+) \tag{22}$$

where,

$$L_{i+1/2}^- = \begin{cases} \tilde{G}(W_i^n,W_{i+1}^n) + \tilde{H}(W_i^n,W_{i+1}^n), \text{if } (\bar{S}_M)_{i+1/2}^n \leq 0 \\ \tilde{G}(W_i^n,W_{i+1}^n), & \text{if } (\bar{S}_M)_{i+1/2}^n > 0 \end{cases}$$

$$L_{i-1/2}^+ = \begin{cases} \tilde{G}(W_{i-1}^n,W_i^n), & \text{if } (\bar{S}_M)_{i-1/2}^n \leq 0 \\ \tilde{G}(W_{i-1}^n,W_i^n) - \tilde{H}(W_{i-1}^n,W_i^n), \text{if } (\bar{S}_M)_{i-1/2}^n > 0 \end{cases}.$$

4 NUMERICAL SIMULATION

There are several classic wave-speed estimating method of HLLC method. One of them is max-min eigenvalues method [7] (23).

$$\bar{S}_L = \min(\bar{u}_L - \bar{c}_L, \bar{u}_R - \bar{c}_R), \bar{S}_R = \max(\bar{u}_L + \bar{c}_L, \bar{u}_R + \bar{c}_R)$$
$$S_L = \min(u_L - c_L, u_R - c_R), S_R = \max(u_L + c_L, u_R + c_R)$$
(23)

Another estimating method is based on Roe average method [8, 9] (24).

$$\bar{S}_L = \bar{u}_{Roe} - \bar{c}_{Roe}, \bar{S}_R = \bar{u}_{Roe} + \bar{c}_{Roe}, S_L = u_{Roe} - c_{Roe}, S_R = u_{Roe} + c_{Roe},$$
(24)

where $\bar{u}_{Roe}$, $\bar{c}_{Roe}$, u_{Roe} and c_{Roe} Roe average velocity and sound [10] of solid phase and gas phase.

The third wave-speed estimating method is pressure-velocity based [6] method (25). Their coefficients $\bar{q}_K$ and q_K are (26), K=L or R.

$$\bar{S}_L = \bar{u}_L - \bar{c}_L \bar{q}_L, \bar{S}_R = \bar{u}_R + \bar{c}_R \bar{q}_R, S_L = u_L - c_L q_L, S_R = u_R - c_R q_R$$
(25)

$$\bar{q}_K = \begin{cases} 1, & \text{if } \bar{p}_K^* \leq \bar{p}_K \\ \sqrt{1 + \dfrac{\bar{\gamma}+1}{2\bar{\gamma}}(\dfrac{\bar{p}_K^* + \bar{p}_\infty}{\bar{p}_K + \bar{p}_\infty} - 1)}, & \text{if } \bar{p}_K^* > \bar{p}_K \end{cases}$$

$$q_K = \begin{cases} 1, & \text{if } p_K^* \leq p_K \\ \sqrt{1 + \dfrac{\gamma+1}{2\gamma}(\dfrac{p_K^*}{p_K} - 1)}, & \text{if } p_K^* > p_K \end{cases}$$
(26)

4.1 Test 1

This problem is a BN model Riemann problem consist of two rarefaction of each phase [6]. The region between the rarefaction waves is close to vacuum; therefore, it is useful to assess the density and pressure positivity in different numerical methods. Both phases are ideal gas with $\bar{\gamma} = \gamma = 1.4$ and $\bar{p}_\infty = 0$. CFL number is 0.4 and mesh number is 400. The domine length is 1, separated by a discontinuity at a position 0.5. Initial data of test 1 is shown in Table 1. We use our method and three other classical methods to simulation the problem. Three other methods to estimate wave-speed are max–min eigenvalues wave-speed, Roe-average eigenvalues wave-speed and pressure-velocity based wave-speed. The wave-speed estimating method based on Roe-average eigenvalues cannot finish the simulation. The other results at time 0.15 are shown in Figure 2. The red line is exact solution. The green circle shows our

Table 1: Initial date of test 1.

	α_L	ρ_L	u_L	p_L	α_R	ρ_R	u_R	p_R
Solid	0.8	1	-2	0.4	0.3	1	2	0.4
Gas	0.2	1	-2	0.4	0.7	1	2	0.4

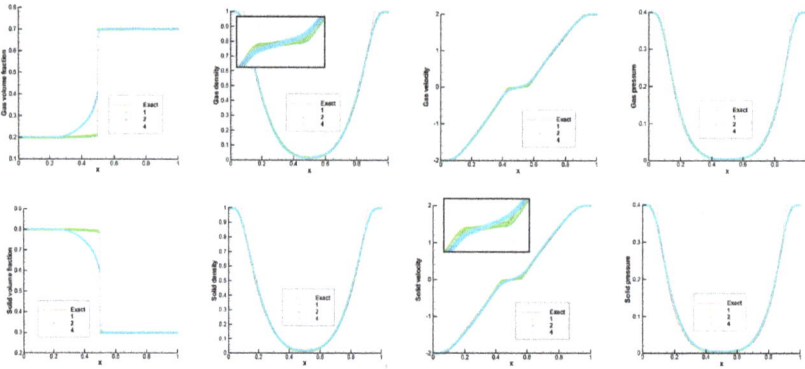

Figure 2: Results of Test 1.

method results. The blue triangle and the light blue square are results of max-min eigenvalues and pressure–velocity based wave-speed estimating method.

The simulation of volume fraction of our method is close to the exact solution, and it is better than the other two methods. What's more, the simulation of velocity of each phase of our method is better than the other two methods.

4.2 Test 2

This problem is a BN model Riemann problem consist of coinciding shocks and rarefactions [3, 11]. The test 2 mainly shows that the waves of each phase are independent each other except near the contact of the solid phase. $\bar{\gamma} = \gamma = 1.4$ and $\bar{p}_\infty = 0$. CFL number is 0.9 and mesh number is 400. The domine length is 1, separated by a discontinuity at a position 0.5. Initial data of test 2 is shown in Table 2. Results of test 2 at time 0.1 are shown in Figure 3. The pressure-velocity based wave-speed estimating method cannot finish the simulation. The red line is exact solution. The green circle shows our method results. The blue triangle and the light blue square are results of max-min eigenvalues and Roe-average wave-speed estimating method. The simulation of solid density of our method is better than the other two methods.

Table 2: Initial data of test 2.

	α_L	ρ_L	u_L	p_L	α_R	ρ_R	u_R	p_R
Solid	0.1	0.2068	1.4166	0.0416	0.2	2.2263	0.9366	6.0
Gas	0.9	0.5806	1.5833	1.375	0.8	0.489	-0.70138	0.986

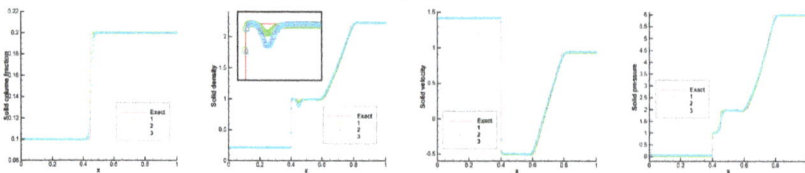

Figure 3: Results of Test 2.

Table 3: Initial data of test 3.

	α_L	ρ_L	u_L	p_L	α_R	ρ_R	u_R	p_R
Solid	0.8	1	0.75	1	0.3	0.125	0	0.1
Gas	0.2	1	0.75	1	0.7	0.125	0	0.1

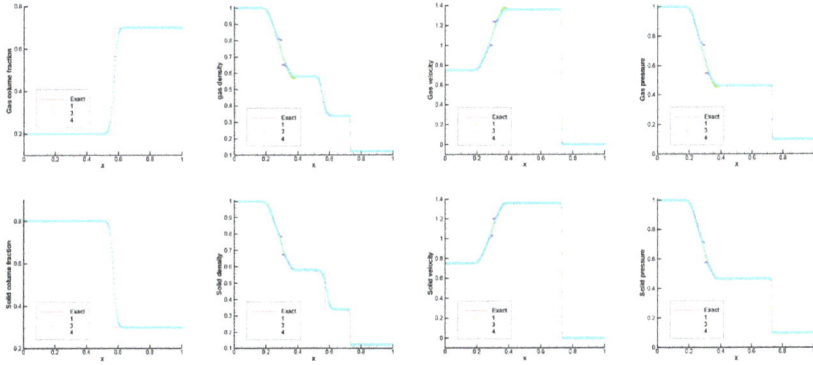

Figure 4: Results of Test 3.

4.3 Test 3

Test 3 is an initial Riemann problem with sonic rarefaction [6]. The initial data of test 3 is shown in Table 3. Both phases are ideal gas with $\bar{\gamma} = \gamma = 1.4$ and $\bar{p}_\infty = 0$. CFL number is 0.9 and mesh number is 400. The domine length is 1, separated by a discontinuity at a position 0.5. Results of test 3 at time 0.2 are shown in Figure 4. The red line is exact solution. The green circle shows our method results. The blue triangle and the light blue square are results of Roe-average wave-speed and pressure–velocity based wave-speed estimating method. Our method need entropy fix for this problem (details are in appendix). If we add entropy fix to our method, The simulation results fits well with exact solution. The simulation of Roe-average wave-speed estimating method is not good for the left sonic rarefaction, which need entropy fix, too. The simulation of pressure–velocity based wave-speed estimating method is well fitting exact solution. The simulation of max-min eigenvalues method appear unphysical oscillation. We make further study and find that it need smaller CFL number.

4.4 Test 4

Test 4 is from [12], with initial data shown in Table 4. $\bar{\gamma} = \gamma = 1.4$ and $\bar{p}_\infty = 0$. CFL number is 0.9 and mesh number is 400. The domine length is 0.06 separated by a discontinuity at a position 0.03 Results of test 3 at time 0.01 are shown in Figure 5. The red line is the exact solution. The green circle shows our method results. The blue triangle, the light blue square and black diamond are results of max–min eigenvalues, Roe-average wave-speed and pressure-velocity based wave-speed estimating method. The velocity of solid phase simulated by our method is better than the other two methods. The simulation of Roe-average wave-speed estimating method cannot simulate the problem correctly. The reason is that the wave-speed

Table 4: Initial data of test 4.

	α_L	ρ_L	u_L	p_L	α_R	ρ_R	u_R	p_R
Solid	0.3	0.553	-0.0553	0.3527	0.7	1.264	-0.115	1.1234
Gas	0.7	5.71	-0.75	6.36	0.3	2.02	0.86	1.87

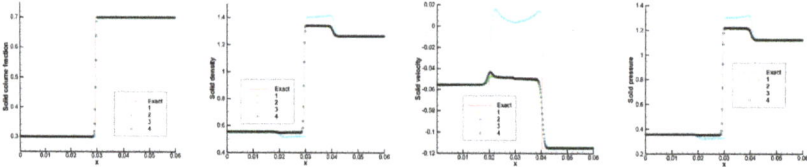

Figure 5: Results of Test 4.

of Roe-average wave-speed estimating method is small, so it cannot catch all characters of the waves.

5 CONCLUSION

Wave-speed is important in simulation of HLLC method. If the wave-speed is too small, the wave characters cannot catch; if it is too large, there is more viscosity in the calculation. There are a lot of methods to estimate the wave-speed, but every method has its applicability. We present a modified HLLC method without estimating wave-speed. It is compared with three classic wave-speed estimating method in the prescript. Numerical results validate the current method for simulation of several typical BN model initial problems. The current method can get the wave-speed directly, so the simulation results are better for different problems.

APPENDIX ABOUT ENTROPY FIX

Nonlinear wave was thought as strong discontinue in HLLC method, so it can be well fitting for contact and shock wave. But for rarefaction wave, if the rarefaction is not transonic (x/t=0 is not in the rarefaction), the simulation of our method is accurate; if the rarefaction is transonic (x/t=0 is in the rarefaction), the simulation need entropy fix.

Now we analyse the left rarefaction, for example. If $\lambda_L = u_L - c_L < 0$ and $\lambda_R = u_L^* - c_L^* > 0$, the ectropy fix is needed. We use the entropy fix similar to Harten-Hyman entropy fix [13], as shown in Figure 6, the discontinue with wave-speed S_L divided into two discontinue at speed λ_L and λ_R. Apply integral form conservation laws, we can get

$$\lambda_L(U^* - U_L) + \lambda_R(U_L^* - U^*) = S_L(U_L^* - U_L)$$

We can get

$$U^* = \frac{(S_L - \lambda_L)U_L + (\lambda_R - S_L)U_L^*}{(\lambda_R - \lambda_L)}$$

Now, flux can be computed as

$$F_{i+1} = F_L + \lambda_L(U^* - U_L) = F_L + \lambda_L \frac{(\lambda_R - S_L)}{(\lambda_R - \lambda_L)}(U_L^* - U_L)$$

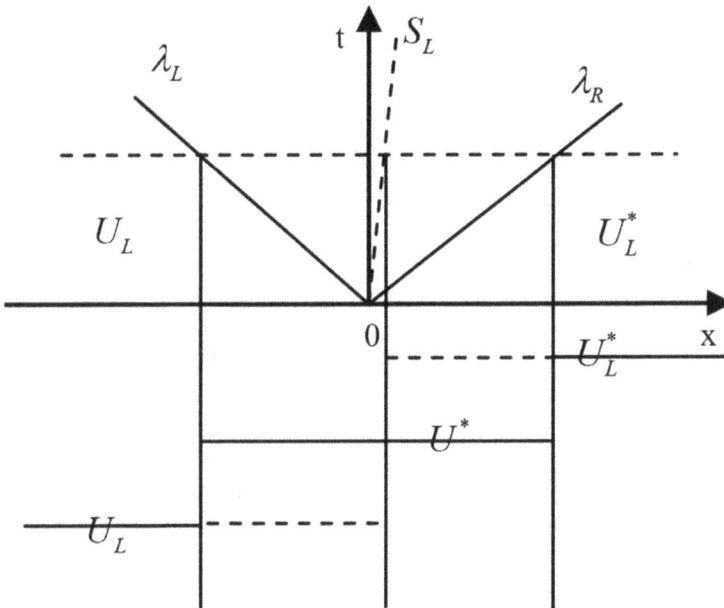

Figure 6: Entropy fix for left rarefaction.

REFERENCES

[1] Baer, M.R. & Nunziato, J.W., A two-phase mixture theory for the deflagration-to-detonation transition in reactive granular materials. *International Journal of Multiphase Flow*, **12**, pp. 861–889, 1986. doi:10.1016/0301-9322(86)90033-9

[2] Embid, P. & Baer, M., Mathematical analysis of a two-phase continuum mixture theory. *Continuum Mechanics and Thermodynamics*, **4**, pp. 279–312, 1992. doi:10.1007/bf01129333

[3] Andrianov, N. & Warnecke, G., The Riemann problem for the Baer-Nunziato two-phase flow model. *Journal of Computational Physics*, **195**, pp. 434–464, 2004. doi:10.1016/j.jcp.2003.10.006

[4] Schwendeman, D.W., Wahle, C.W. & Kapila, A.K., The Riemann problem and a high-resolution Godunov method for a model of compressible two-phase flow. *Journal of Computational Physics*, **212**, pp. 490–526, 2006. doi:10.1016/j.jcp.2005.07.012

[5] Deledicque, V. & Papalexandris, M.V., An exact Riemann solver for compressible two-phase flow models containing non-conservative products. *Journal of Computational Physics*, **222**, pp. 217–245, 2007. doi:10.1016/j.jcp.2006.07.025

[6] Tokareva, S.A. & Toro, E.F., HLLC-Type Riemann solver for the Baer-Nunziato compressible two-phase flow. *Journal of Computational Physics*, **229**, pp. 3573–3604, 2010. doi:10.1016/j.jcp.2010.01.016

[7] Davis,S.F., Simplified second-order Godunov-type methods. *SIAM Journal on Scientific and Statistical Computing*, **9**, pp. 445–473, 1988. doi:10.1137/0909030

[8] Davis, S.F., Simplifed second-order Godunov type methods. *SIAM Journal on Scientific and Statistical Computing*, **9**, pp. 445–473, 1988. doi:10.1137/0909030

[9] Einfeldt, B., On Godunov-type methods for gas dynamics. *SIAM Journal on Numerical Analysis*, **25**(?), pp. 294–318, 1988. doi:10.1137/0725021

[10] Roe,P.L. & Pike, J., Efficient construction and utilisation of approximate Riemann solutions. In *computing methods in applied science and engineering*. North-Holland, 1984.

[11] Pan, L., Zhao, G., Tian, B. & Wang, S., A gas kinetic scheme for the Baer-Nunziato two-phase flow model. *Journal of Computational Physics*, **231**, pp. 7518–7536, 2012. doi:10.1016/j.jcp.2012.04.049

[12] Lowe, C.A., Two-phase shock-tube problems and numerical methods of solution [J]. *Journal of Computational Physics*, **204**, pp. 598–632, 2005. doi:10.1016/j.jcp.2004.10.023

[13] Harten, A. & Hyman, J.M., Self adjusting grid methods for one-dimensional hyperbolic conservation laws. *Journal of Computational Physics*, **50**, pp. 235–269, 1983. doi:10.1016/0021-9991(83)90066-9

[14] Harten, A., Lax, P.D. & van Leer, B., On upstream differencing and Godunov-type schemes for hyperbolic conservation laws. *SIAM Review*, **25**(1), pp. 35–61, 1983. doi:10.1137/1025002

[15] Toro, E.F., Spruce, M. & Speares, W., Restoration of the contact surface in the HLL-Riemann solver. *Shock Waves*, **4**, pp. 25–34, 1994. doi:10.1007/bf01414629

[16] Toro, E.F., *Riemann solvers and numerical methods for fluid dynamics*, Berlin: Springer, 1999.

EFFECT OF GROOVE ANGLE AND SEPARATION ON THE MICRO-JET FORMS

L. CHAO, W. PEI, F. QIJING
Institute of Applied Physics and Computational Mathematics, Beijing, China.

ABSTRACT

When a shock wave propagates through a metal surface with groove defects, micro-jet can be emitted from the free surface of the sample. The mass, velocity and distribution of the micro-jet varies depended not only on the initial shock conditions and material properties of the metal sample, but also on the groove angle and separation on the sample surface. To understand the effect of groove angle and separation on the micro-jet forms, a elastic plastic hydrodynamics Eulerian code was applied in the simulation. The maximum velocity, ejecting factor and mass–velocity distribution of the micro-jet from different groove angle and separation was presented. The simulation results show: the maximum velocity decreased linearly as the groove angle increased, but was insensitive to the change of groove separation. The ejecting factor and mass–velocity distribution changed remarkably when the groove angle and separation changed. But when the groove separation was longer than five times of the groove depth, the ejecting factor and mass–velocity distribution of the micro-jet maintained constant.

Keywords: Euler hydrodynamic method; groove angle; micro-jet; shock loading

1 INTRODUCTION

As a shock induced surface effect, the ejection was known as the amount of material traveled faster than free surface, when the shock wave reflected from it. The machine marks, grain boundaries, inclusions, material voids, and damaged layers, they all could be the possible sources that contributed to the formation of ejecta [1].

Since 1953 the ejecting phenomenon was observed [2], many experiment and analysis researches have been carried out. Mass ejection experiments indicated that micro-jet from defect on the specimen surface was the main source of ejecta when the surface melting didn't occur [3]. The ejection experiment conducted on groove specimen showed that the groove angle effected the maximum ejecting velocity and total ejecting mass [4]. Mass ejection experiments conducted on Sn indicated that the surface roughness had great effect on the ejecting mass [5]. The MD simulation result showed that average velocity of the ejecting particles increased with the groove angle, and the amount of ejecting particles decreased as groove angle increased, Jun [6]. The SPH simulation result indicated that the ejecting factor reached its maximum as the half groove angle equaled to $45°$, and the maximum velocity of ejecta showed a linear reduction with the increase of groove angle, Pei [7]. The experimental and numerical results indicated that the free surface defect was an important source of ejecta, but some of the results weren't consistent, therefore the defect induced ejection needed to be investigated thoroughly.

There were two kinds of specimens used in mass ejection experiment. One was machined metal surface with nature defects of periodic V-groove. The other was specimens with artificial defects of parallel V-groove, which was used to study the groove angle effect in some ejecting experiments. The separation and depth of the two kinds of V-groove were different. Those might contribute to the diverse trends in the experimental and numerical simulation result. A Euler code was used in simulating the two kinds of mass ejection models. When the groove angle and the groove separation changed, the maximum ejecting velocity and the mass distribution of micro-jet were focused on.

2 COMPUTATIONAL METHOD

A elastic plastic hydrodynamic L-R type two steps Euler method Meph was applied in simulating the micro-ejection from the groove of free surface. Meph was appropriated to simulate the high velocity impact and large deformation problems, Qijing [8, 9]. A simplified equation of state for condensed medium and an elastic-perfectly plastic constitutive model were used, see eqns (1) and (2).

Equation of state:

$$p = C_0^2 (\rho - \rho_0) + (\gamma - 1) \rho e \qquad (1)$$

- ρ_0 is initial density.
- C_0 is sound velocity.
- γ is Constant of Material.
- ρ is the continuum density.
- e is the specific internal energy

Constitutive relationship:

$$S_{xx}^2 + S_{yy}^2 + S_{zz}^2 + 2(S_{xy}^2 + S_{yz}^2 + S_{zx}^2) \leq \frac{2}{3}(Y_0)^2 \qquad (2)$$

- S_{ij} is deviatoric stress.
- Y_0 is yield strength.

3 COMPUTATIONAL MODEL

There were two kinds of specimens using in mass ejection experiment: (a) Machined metal surface with nature defects of periodic V-groove; (b) in order to study the groove angle effect, specimens with artificial defects of parallel V-groove were used in some ejecting experiment, see Figure 1. The study was focused on the maximum ejecting velocity and the mass distribution of micro-jet, when the groove angle and the groove separation changed.

The specimens used in the experiment were conducted on aluminum with artificial defects of parallel V-groove, Asay [4]. The groove depth was 55um, the groove separation was 130 um, the half groove angle was vary from 15° to 45°. Two kinds of models were used in the

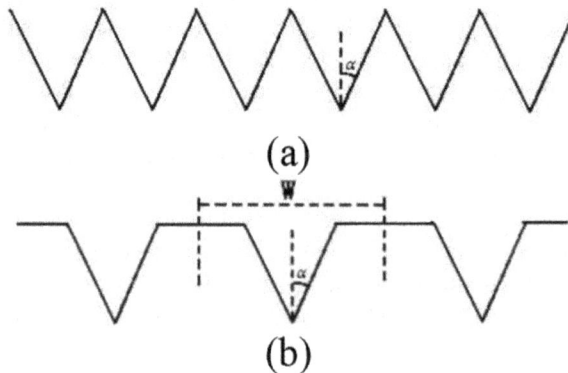

Figure 1: Idealized defects on two kinds of surfaces. (a) Machined metal surface; (b) Artificial grooved surface.

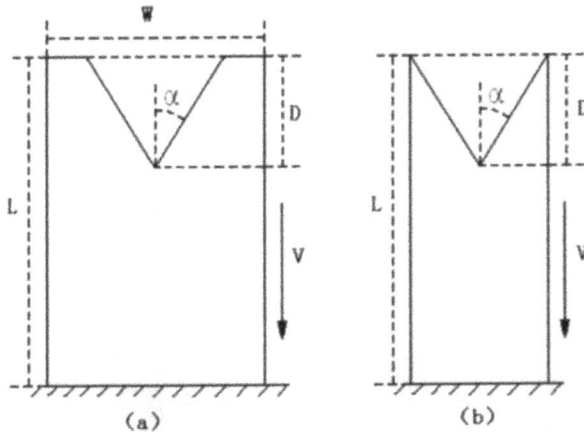

Figure 2: Numerical simulation model. (a) Model 1; (b) Model 2.

numerical simulation, see Figure 2. Model (a), the groove depth was 55 um, the groove separation was 130 um, the half groove angle vary from 15° to 50°, the height of model was 200 um. Model (b), the groove depth and the height of model were same as model a, the half groove angle was vary from 15° to 75°, the groove separation varied with the angle.

4 SIMULATION RESULT

4.1 The micro-jetting forming process

At the beginning, model (a) impacted rigid boundary at 1.5 km/s, produced a Al shock pressure of about 30 GPa, see Figure 3(a). When the shock wave arrived the bottom of groove, the rarefaction wave that reflected from the groove sidewall accelerated the material of the groove

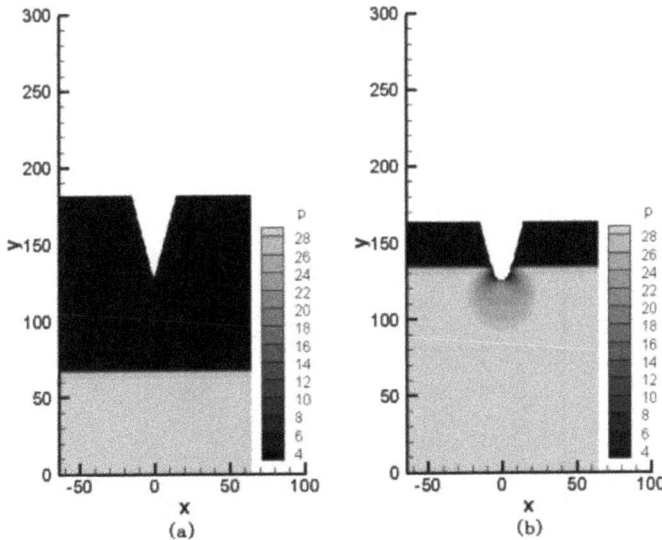

Figure 3: Pressure isoline imagine.

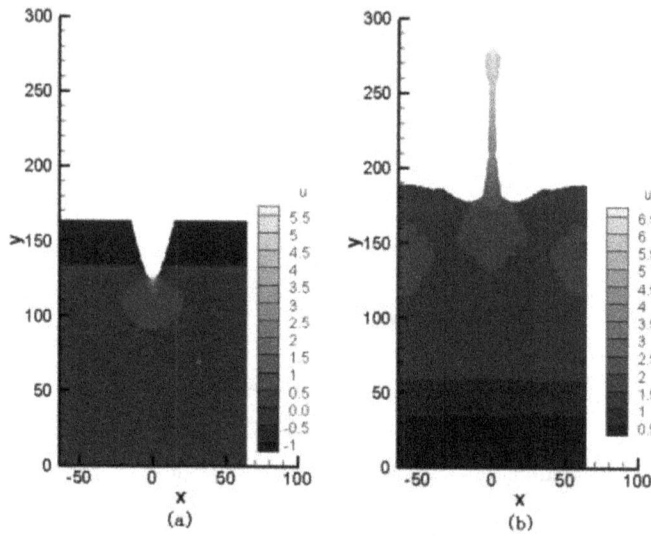

Figure 4: Velocity isoline imagine.

bottom, see Figures 3(b) and 4(a). Groove sidewalls collapsed under the combined effect of shock wave and rarefaction wave, the fragmentation impacted on symmetry axis. Under the high pressure of collision, a high-speed jet formed, see Figure 4(b).

4.2 Effect of groove angle

The maximum velocity of different angle grooves was given, see Figure 5. Where V_{max} was the maximum velocity of micro-jetting, V_{fs} was the free surface velocity. The simulation result of two kinds of model exposed that maximum velocity was fitting well with the

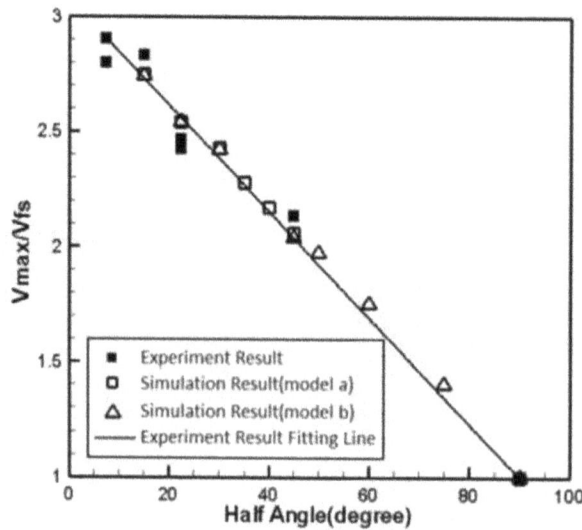

Figure 5: Maximum velocity of different angle grooves,

experiment data. The maximum velocity decreased linearly as the groove angle increased. The maximum velocity was insensitive to the change of groove separation.

The ejecting factor (ejecting mass divided by the product of groove volume and initial material density) simulation result of model (a) fitted well with the experiment data, as shown in Figure 6. The simulation result of model (a) and experiment data indicated that the ejecting factor decreased as the groove angle increased, when the groove separation was fixed. For model (b), the ejecting factor was not monotone; the maximum of ejecting factor was acquired when the half groove angle was about 60°.

Compared with model (a) whose groove separation was fixed, the simulation result of small groove angle cases in model (b) was slimmer. That meant for the same groove angle, the longer groove separation was, the more ejecting mass.

The mass and velocity distribution curves of the two kinds of models were given in Figure 7, where V was micro-jetting velocity, V_{fs} was the free surface velocity. Y-axis was the ejecta

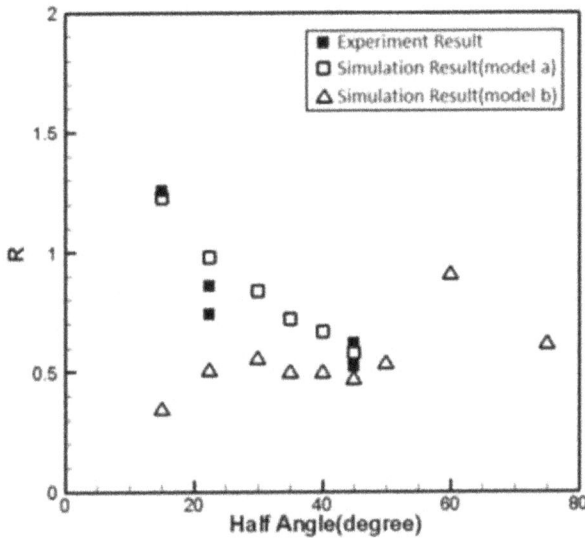

Figure 6: Ejecting factor of different angle grooves.

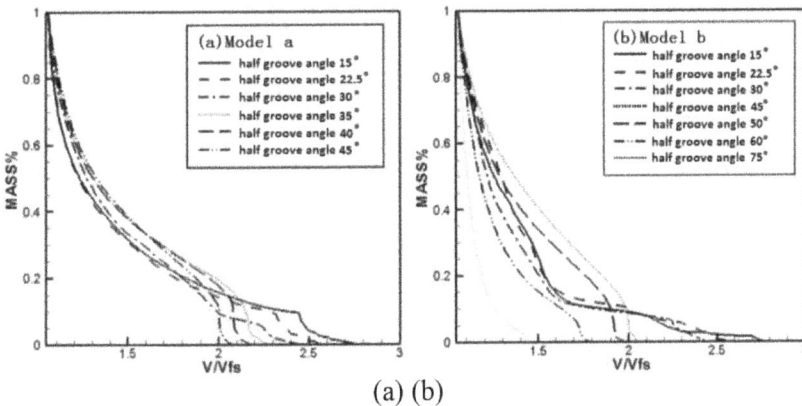

(a) (b)

Figure 7: M–V distribution curve for ejection of different groove angles.

mass, whose non-dimensional velocity was more than the corresponding abscissa value, divided by total ejecting mass.

The simulation results of model (a) showed that the smaller the groove angle, the more the ratio of micro-jet tail mass (whose velocity less than 1.1 times of free surface velocity) to the total jet mass was. The simulation results of model (b) indicated that the bigger the groove angle was, the less ratio of micro-jet tail mass to total jet mass, when the groove angle was less than 45°; the smaller groove angle was, the less ratio of micro-jet tail mass to total jet mass, when the groove angle was bigger than 45°. For models (a) and (b) the smaller groove angle was, the faster maximum velocity of micro-jetting.

4.3 Effect of groove separation

Ejecting factor of different groove separation was given in Figure 8, DIS was non-dimensional distance (separation divided by groove depth). It was concluded that the smaller groove angle, the bigger ejecting factor was, when the groove separation was fixed. For the three cases in Figure 8, when the groove separation was smaller than four times of groove depth, the ejecting factor increased fast with the increasing of separation; when the groove separation was bigger than four times but less than five times of groove depth, the ejecting factor increased slowly with the increase of separation; when the groove separation was bigger than five times of groove depth, the total ejecting mass or the ejecting factor remained constant.

The statistical result in Figure 9, the M–V distribution curve of different groove separation showed that velocity of micro-jet head was insensitive to the change of groove separation. But the ratio of micro-jet head mass to total jet mass decreased with the increase of groove separation. The ratio of micro-jet tail mass to total jet mass increased with the increasing of groove separation. The M–V distribution curve of ejecta changed very minim, when the groove separation was bigger than 4.4 times of groove depth. The total ejecting mass remained constant, and the M–V distribution curve was invariable, when the groove separation was bigger than five times of groove depth.

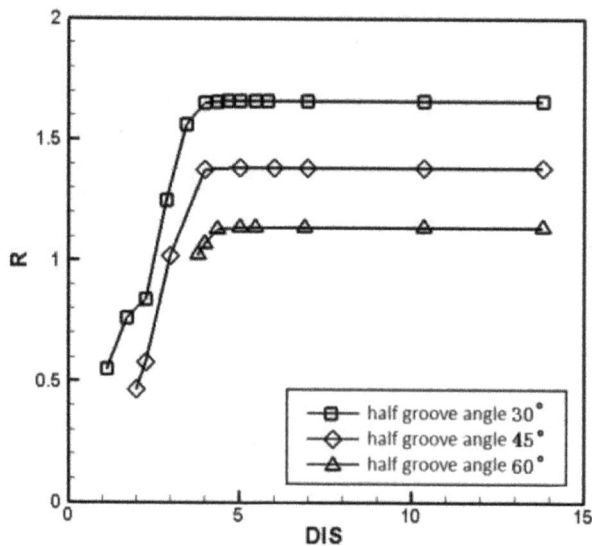

Figure 8: Ejection factor of different groove separation.

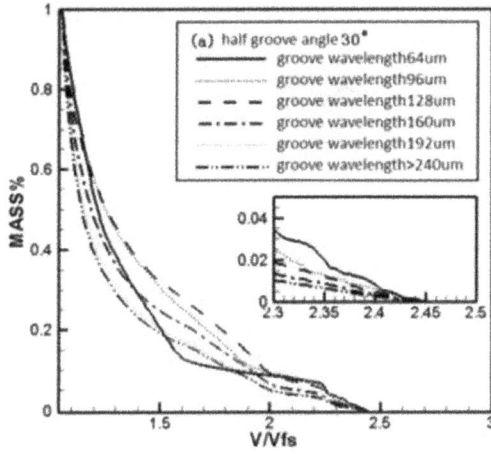

(a)

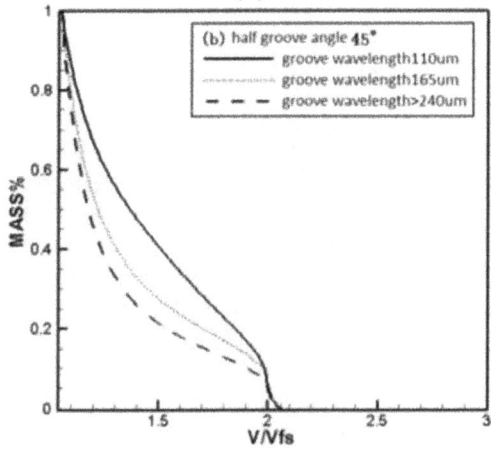

(b)

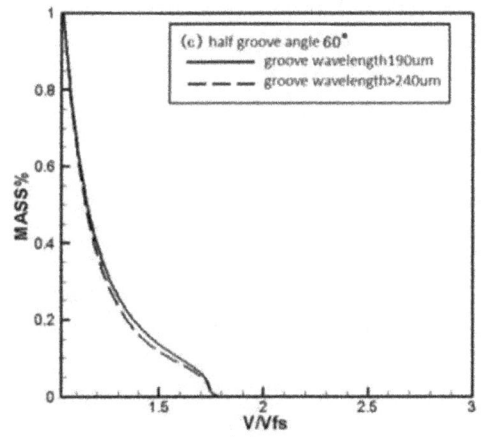

(c)

Figure 9: M–V distribution curve for ejection of different groove separation.

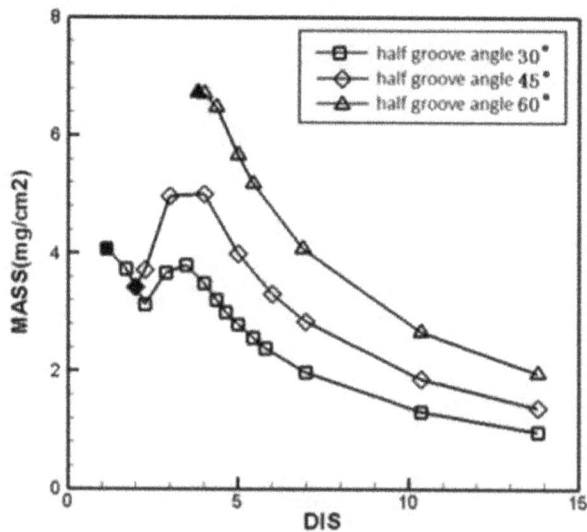

Figure 10: Ejecta mass of different groove separation.

In the ejection experiment, Asay foil or quartz was usually used to obtain the ejecta mass from unit surface of specimen. In order to study the effect of groove separation on the ejecting mass from unit surface, the ejecting mass of different groove separation was given in Figure 10. The Y-axis was total mass of micro-jet per square centimeter from the specimen surface. The hollow square, diamond and triangle represented the simulation results of 60°, 45° and 30° V-grooves, respectively. The solid one was the corresponding experiment result.

It was showed in Figure 10, when the groove separation was less than five times of groove depth, the ejecting factor increased, and the number of grooves in the unit surface decreased with the increasing of groove separation. Therefore ejecting mass from unit specimen surface was non-monotone with the increase of groove separation, when the groove separation was less than five times of groove depth. When the groove separation was more than five times of groove depth, the ejecting factors remained a constant, but the number of grooves in the unit surface decreased, so the ejecting mass from unit specimen surface was decreased as the groove separation increased.

5 CONCLUSION

A elastic plastic hydrodynamics Euler code was applied to simulate the micro-jet, which formed from the grooves in the surface under shock loading. The study was focused on the velocity and the mass of micro-jet, when the groove angle and the groove separation changed.

The simulation results showed that when the groove angle increased, the maximum velocity decreased linearly, the ejecting factor and mass-velocity distribution of the micro-jet changed obviously. The maximum velocity was insensitive to the change of groove separation, but the ejecting factor and mass-velocity distribution changed remarkably when the groove separation increased. When the groove separation was more than five times of the groove depth, the ejecting factor and mass–velocity distribution of the micro-jet maintained invariant, and the ejecting mass from unit specimen surface was decreased with the groove separation increasing.

REFERENCES

[1] Sorenson, D.S., Minich, R.W., Romero, J.L., Tunnell, T.W. & Malone, R.M., Ejecta particle size distributions for shock loaded Sn and Al metals. *Journal of Applied Physics*, **92**(10), pp. 5830–5836, 2002. doi: 10.1063/1.1515125

[2] Walsh, J.M., Shreffler, R.G. & Willing, F.J., Limiting conditions for jet formation in high velocity collisions. *Journal of Applied Physics*, **24**(3), pp. 349–359, 1953.

[3] Asay, J.R., *A model for estimating the effects of surface roughness on mass ejection from shocked materials. SAND78-1256.* Sandia Laboratories, 1978.

[4] Asay, J.R., *Material ejection from shock-loaded free surface of aluminum and lead.* SAND76-0542. Sandia Laboratories,1976.

[5] Zellner, M.B., Grover, M., Hammerberg, J.E., Hixson, R.S., Iverson, A.J., Macrum, G.S., Morley, K.B., Obst, A.W., Olson, R.T., Payton, J.R., Rigg, P.A., Routley, N., Stevens, G.D., Turley, W.D., Veeser, L. & Buttler, W.T., Effects of shock-breakout pressure on ejection of micron-scale material from shocked tin surfaces. *Journal of Applied Physics*, **102**(013522), 2007. doi:10.1063/1.2752130

[6] Jun, C., Fuqian, J. & Jinglin, Z., Molecular dynamics simulation of micro particle ejection from a shock-impacted metal surface. *Acta Physica Sinica*, **51**(10), pp. 2386–2392, 2002.

[7] Pei, W., Jianli, S. & Chengsen, Q., Study of groove angle effect on micro-jet from shocked metal surface. *Acta Physica Sinica*, **61**(23), 234701, 2012.

[8] Qijing, F., Pengcheng, H., Yihong, H., Eulerian numerical simulation of a shaped charge. *Explosion and Shock Waves*, **28**(2), pp. 138–143, 2008.

[9] Qijing, F., Xianlin, C., Pengcheng, H., Fengguo, Z., Shudao, Z., Quanmin, L. & Haibo, G., The simulating method study and application of well perforator operating process. *Chinese Journal of Computational Physics*, **26**(6), pp. 887–891, 2009.

Author index

Complex Systems

Theory and Applications

Edited by: **G. RZEVSKI**, *The Open University, UK and* **C.A. BREBBIA**, *Wessex Institute, UK*

This multi-disciplinary book presents new approaches for resolving complex issues that cannot be resolved using conventional mathematical or software models.

Complex Systems occur in an infinite variety of problems encompassing fields as diverse as economics, the environment, humanities, social and political sciences, physical sciences and engineering.

The papers in the book cover such topics as: Complex business processes; Supply chain complexity; Complex adaptive software; Management of complexity; Complexity in social systems; Complexity in engineering; Complex issues in biological and medical sciences; Complex energy systems Complexity and evolution.

ISBN: 978-1-78466-235-6   eISBN: 978-1-78466-236-3
Published 2017 / 260pp